SPECTACULAR FLOPS
Game–Changing
Technologies That Failed

SPECTACULAR FLOPS
Game–Changing
Technologies That Failed

MICHAEL BRIAN SCHIFFER
Smithsonian Institution
Washington, D.C.

ELIOT WERNER PUBLICATIONS, INC.
CLINTON CORNERS, NEW YORK

Library of Congress Cataloging-in-Publication Data

Names: Schiffer, Michael B., author.
Title: Spectacular flops : game-changing technologies that failed /
  Michael Brian Schiffer.
Description: Clinton Corners, New York : Eliot Werner Publications,
  [2019] | Includes bibliographical references and index.
Identifiers: LCCN 2018044745| ISBN 9780989824996 |
  ISBN 0989824993
Subjects: LCSH: Inventions – History – Popular works. | New
  Products – History – Popular works. | System failures
  (Engineering) – History – Popular works. | Errors – History –
  Popular works. | Business failures – History – Popular works.
Classification: LCC T47 .S35 2019 | DDC 509 – dc23
LC record available at https://lccn.loc.gov/2018044745

ISBN-10: 0-9898249-9-3
ISBN-13: 978-0-9898249-9-6

Copyright © 2019 Eliot Werner Publications, Inc.
PO Box 268, Clinton Corners, New York 12514
http://www.eliotwerner.com

Printed in the United States of America

PREFACE

I first heard the term "spectacular flop" around 1980, when my friend Bill Rathje and I had just completed a draft of our unconventional—even radical—textbook *Archaeology*. Our acquisitions editor at Harcourt Brace Jovanovich, Peter Dougherty, liked our innovative approach but worried that instead of a stunning success it might become a spectacular flop. His concern was prescient, as the book underperformed sales projections by more than fifty thousand copies and did not go into a second edition. *Archaeology* was, indeed, a spectacular flop.

The history of technology teaches us that many ambitious technologies failed to meet their promoters' lofty expectations. This book presents a sample of spectacular flops from the late 18th to the early 21st centuries. The case studies are biased toward my interests: transportation, public works, and energy-related technologies, from the world's first automobile to nuclear fusion reactors. Along the way the reader, who is assumed to have no previous familiarity with the technologies, is introduced to basic principles of (for example) steam power, electricity, and jet engines.

When my colleagues learned I was working on this book, they wondered—knowing my propensity to theorize about all things technological—if I would be able to confect any generalizations about spectacular flops. I too had my doubts because the category encompasses technologies of great variety that flopped from diverse causes; perhaps no generalizations could encompass them all. And I found it so, for much cogitation failed to yield consequential, pan-technology generalizations. Reluctant to abandon the quest, however, I was able to formulate a handful of limited, carefully qualified generalizations, tucked into Chapter 14.

I thank the "document delivery" staff of the University of Arizona Library for easing my research burdens by at least an order of magnitude. I am grateful to the Lemelson Center for the Study of Invention and Innovation at the National Museum of American History (NMAH) for providing me with an institutional home. I am indebted to Smithsonian curators Deborah J. Warner and Peggy Kidwell for their helpful suggestions and insightful comments on ideas I shared with them during our Tuesday lunches at NMAH.

Eliot Werner did a superb job of shepherding *Spectacular Flops* through the production process; I have never worked with a more capable editor. As always, my wife Annette has been a sounding board for ideas and is a constant source of love and support. I could never complete these writing projects without her encouragement and helpful advice.

CONTENTS

Chapter 1 • Introduction: Understanding Technological Failures / 1

Chapter 2 • Secret Project: Nicolas-Joseph Cugnot and the First Automobile / 13

Chapter 3 • Too Little, Too Late: The Atmospheric Railway / 33

Chapter 4 • Putting Electromagnetism to Work / 55

Chapter 5 • Audacious Engineer: Isambard Kingdom Brunel and the *Great Eastern* Steamship / 71

Chapter 6 • Ferdinand de Lesseps' Panama Canal / 95

Chapter 7 • Crackpot Invention? Nikola Tesla's World System / 115

Chapter 8 • Visionary Inventor: R. Buckminster Fuller's Dymaxion World / 135

Chapter 9 • The Nuclear-Powered Bomber / 157

Chapter 10 • Domesticating the Bomb: "Geographical Engineering" and Project Chariot / 177

Chapter 11 • Chrysler's Turbojet-Powered Automobile / 197

Chapter 12 • The Concorde: Supersonic Airliner / 219

Chapter 13 • Fusion, Hot and Cold / 239
Chapter 14 • Conclusion: Several Limited
      Generalizations / 259
References / 275
Index / 295

FIGURES

2.1 Heron's jet-powered philosophical device. Source: Thurston (1891:Frontispiece).

2.2 Thomas Savery's condensing steam engine (1698). Source: Thurston (1891:37).

2.3 Thomas Newcomen's condensing steam engine (1712). Source: Galloway (1881:Plate I).

2.4 Jacob Leupold's design for an expansion steam engine. Source: Leupold (1724:Table XLIII, Figure II). Courtesy of Dibner Library, Smithsonian Institution.

2.5 Nicolas-Joseph Cugnot's *fardier*. Source: Figuier (1891:Figure 123).

2.6 Driving mechanism of Cugnot's *fardier*. Source: author's photograph.

2.7 Alain Cerf's replica of Cugnot's *fardier*. Source: courtesy of the Tampa Bay Automobile Museum.

3.1 George Stephenson's rocket (ca. 1829). Source: Findlay (1889:XVII).

3.2 Details of the atmospheric railway. Source: Heck (1851:429).

3.3. The South Devon atmospheric railway. Source: a painting by Nicholas Condy (1848). Note Dawlish pumping station at right.

4.1 Joseph Henry's electric teeter-totter device. Source: Henry (1831:342).

4.2 Patent model of Thomas Davenport's electric motor. Source: National Museum of Natural History, Smithsonian Institution, author's photograph.

4.3 Charles Page's electric locomotive. Source: *Greenough's American Polytechnic Journal* 4(1854):257.

5.1 The *Great Western*. Source: Nicholls and Taylor (1882:311).

5.2 The *Great Britain*. Source: Claxton (1845:Frontispiece).

5.3a The *Great Eastern*'s transverse bulkheads under construction. Source: courtesy of National Museum of American History, Smithsonian Institution, photograph by Robert Howlett.

5.3b The *Great Eastern*'s double hull under construction. Source: courtesy of National Museum of American History, Smithsonian Institution, photograph by Robert Howlett.

5.4a The *Great Eastern*'s paddle wheel engine. Source: author's collection from the *Illustrated London News*, ca. 1858.

5.4b The *Great Eastern*'s screw engine. Source: author's collection from the *Illustrated London News*, ca. 1858.

5.5 Brunel directing the launch of the *Great Eastern*. From left to right: John Scott Russell, Henry Wakefield (Brunel's assistant), Isambard Kingdom Brunel, and Lord Derby. Source: courtesy of Wikimedia Commons; photo by Robert Howlett.

5.6 The *Great Eastern* begins its inaugural voyage. Source: *Illustrated London News*, 23 June 1860, pp. 600–601.

6.1 Map of the Panama Railroad. Source: Otis (1862:v).

6.2 Dredge at work on the Suez Canal. Source: courtesy of Wikimedia Commons, from Tropenmuseum, part of the National Museum of World Cultures.

6.3 French steam-powered bucket excavator at work on the Panama Canal, ca. 1887. Source: *Popular Science Monthly*, December 1887, p. 156.

6.4 Culebra Cut between the two highest hills, 1914. Source: courtesy of the Library of Congress.

7.1   Tesla graces the cover of *Time*. Source: *Time*, 20 July 1931.

7.2   DC electric motor. Source: adapted from Wikimedia Commons, creator: Nokia123~commonswiki.

7.3   An early Tesla induction (AC) electric motor. Source: courtesy of Wikimedia Commons, photograph by Stas Kozlovsky in the Nikola Tesla Museum, Belgrade, Serbia.

7.4   Tesla's Colorado Springs laboratory. Source: courtesy of Wikimedia Commons, Dickenson V. Alley, photographer.

7.5   Tesla's Wardenclyffe laboratory and transmitting tower. Source: Reeve (1911:97).

8.1   Fuller and the model of the 1929 Dymaxion house. Source: by permission of the Estate of R. Buckminster Fuller.

8.2   Fuller shows off a Dymaxion car at the Chicago World's Fair (ca. 1934). Source: by permission of the Estate of R. Buckminster Fuller.

8.3   A Dymaxion deployment unit (1941). Source: courtesy of the Library of Congress.

8.4   Biosphère Montréal. Source: courtesy of Wikimedia Commons, Cédric Thévenet, photographer.

8.5   Fuller and his inventions on a 2004 U.S. postage stamp. Source: author's photograph.

8.6   Wichita house in the Henry Ford Museum. Source: courtesy of Wikimedia Commons, Michael Barera Photographer.

9.1   HTRE-1 and crew. Source: courtesy of the National Reactor Testing Station.

9.2   Cutaway drawing of General Electric's XNJ140E nuclear propulsion system. Source: General Electric (1962: Figure 4.1).

9.3   Artist's conception of Convair's NX2 nuclear bomber. Source: General Electric (1962: Figure 1.2).

10.1   Draft plan for Project Chariot. Source: courtesy of Lawrence Livermore National Laboratory.

10.2   Aerial image showing damage to Ogotoruk Creek area. Source: courtesy of Wikimedia Commons, photographer Rtones, July 2014.

11.1 General Motors Firebird I concept car, 1953. Source: courtesy of Wikimedia Commons, photographer Karrmann.

11.2 Operation of Chrysler's fourth-generation turbine engine. Source: Roy, Hagen, and Belleau (1964:3).

11.3 Chrysler turbine's variable nozzle system. Source: Chrysler Corporation (1963:7).

11.4 Two views of the 1963 Chrysler Turbine. Source: courtesy of the Transportation Division, National Museum of American History, Smithsonian Institution.

11.5 Cutaway view of Chrysler's fourth-generation turbine engine. Source: Chrysler Corporation (1963:2). Key: A = accessory drive; B = compressor; C = right regenerator rotor; D = variable nozzle unit; E = power turbine; F = reduction gear; G = left regenerator rotor; H = gas generator turbine; I = burner; J = fuel nozzle; K = igniter; L = starter-generator; M = regenerator drive shaft; N = ignition unit.

12.1 Concorde's last flight, 26 November 2003, overflying Filton Airfield. Source: courtesy of Wikimedia Commons, photographer Arpingstone.

12.2 Russia's supersonic Tu-144LL taking off at Zhukovsky Air Development Center (1997). Source: courtesy of NASA.

12.3 NASA drawing of nexgen SST. Source: courtesy of NASA.

13.1 Stanley Pons and Martin Fleischmann show off their cold fusion apparatus. Source: courtesy of Special Collections, J. Willard Marriott Library, University of Utah.

13.2 Cutaway drawing of the ITER tokamak. Source: courtesy of ITER.

13.3 Cutaway drawing of the ITER tokamak showing its infrastructure. Source: courtesy of ITER.

SPECTACULAR FLOPS
Game–Changing
Technologies That Failed

1

INTRODUCTION

Understanding Technological Failures

In the early 1990s, Thomas Park Hughes—an eminent historian of technology—visited the University of Arizona where I worked. During those years I co-taught "The Cultural Context of Materials Technology" with W. David Kingery, a distinguished materials scientist. Hughes visited our class and later that day came by my office. During our brief conversation, he asked a question that surprised me. "What," he said, "could an archaeologist contribute to the history of technology?" I replied that only archaeology furnishes evidence for writing the early history (actually *prehistory*) of materials technologies as varied as ceramics, plaster, and copper metallurgy. Acknowledging these significant contributions, Hughes then asked, "What else can archaeology do?"

I mentioned to Hughes my research on the earliest electric automobiles. He showed interest in the project and cautiously accepted my claim that an archaeological approach might offer new insights for addressing longstanding historical issues, particularly why the early electric car was such a short-lived consumer product. At the time of his visit, however, the study was incomplete.

Publications resulting from the electric car project showed that earlier narratives about this era (ca. 1895–1920) were deeply flawed because their explanations relied on folk theories that mainly implicated shortcomings in batteries.[1] Instead my narrative showed that the causes of the early electric car's short lifespan included the patriarchal structure of middle-class American families. Those families had privileged the leisure activities of men (touring in the countryside) over the social and utilitarian activities of women (traveling in town) in their choice of gasoline-powered cars over battery-powered cars. I was gratified that two later studies of early electric vehicles built upon my work and also pointed to social factors.[2]

✳ ✳ ✳ ✳ ✳

My explanation was founded on an evolving behavioral framework for studying technological change. Applying this framework enables the construction of historical narratives that in many cases challenge prevailing ideas. One recent example comes from 19th century electrical technologies.[3] In previous accounts Englishman Michael Faraday is credited with inventing the electric motor. From the primary literature of the early 19th century, I learned that in 1822 Faraday showed that the circular magnetic field around a current-carrying wire can produce circular motion. His elegant experiments advanced electromagnetic science but did not initiate the technological tradition of making motors.

The earliest electric motors—and even many of today—operate on an entirely different principle invented by Joseph Henry, first secretary of the Smithsonian Institution. In 1831 Henry showed that electromagnetism can create continuous mechanical motion by using the motor's own motion to switch an electromagnet's polarity (look ahead to Figure 4.1). Not surprisingly, Henry's invention of this operating principle sired the technological tradition of making motors. This is just one of many examples in which the behavioral framework has helped researchers construct new narratives of technological change.[4]

Clearly, archaeology—prehistoric, historical, industrial, and contemporary—furnishes fresh evidence of ancient and modern technologies, corrects earlier narratives and interpretations, and crafts new narratives informed by an abiding respect for the concrete interactions among people and technologies in their societal contexts. But in reiterating Hughes's question we may ask, "What else does archaeology do in its engagement with historical technologies?" The answer is that we create generalizations—constructs, models, law-like statements, and even theories—about technological change.

Behavioral archaeologists have shown that historical technologies furnish a diverse and essentially unlimited database for originating and evaluating generalizations. After all, historical accounts may contain useful descriptions of both technologies *and* the activities in which they took part. By interrogating historical cases, we are in an advantageous position to formulate generalizations.

For many decades behavioral archaeologists have relentlessly pursued the goal of creating generalizations and heuristics that apply to technological change in any society, past or present.[5] In *Studying Technological Change: A Behavioral Approach*, I fashion an overarching framework that encompasses the many new contributions.[6] Although this book strives to be comprehensive, it pays insufficient attention to a topic of growing interest: technological failures.[7] The present book partially remedies this deficiency by offering generalizations about "spectacular flops," the most conspicuous examples of failed technologies.

A spectacular flop begins life as someone's vision of an ambitious, perhaps audacious, technology that might be. The vision's promoters boldly forecast that it will have a revolutionary impact on individuals, families, companies, industries, or government agencies. At some point, however, the technology's life comes to a premature end, leaving behind great disappointment and (sometimes) great financial losses and tarnished reputations. The failure of such a technology renders it a spectacular flop.

Spectacular flops are a special species of failure because the technologies *usually* (a) involve a multi-year development process and (b) require a vast investment of resources—human, material, societal, and so forth. As just one example, the U.S. government's nuclear-powered bomber project endured from the late 1940s until 1961 and cost over $1 billion (Chapter 9). In addition, (c) spectacular flops often have several contributing causes.

As singular occurrences with multiple causes, spectacular flops seemingly defy attempts at generalization. Admittedly, no single generalization is apt to account for all—or even many—examples. Fortunately, by following the archaeological imperative to seek patterns in data, I have found that some causes (even multiple causes) recur among the case studies in Chapters 2–13. These patterns form the basis of the limited generalizations presented in Chapter 14, which together permit us to understand many failures of the spectacular sort. As just one example, the case of the nuclear bomber suggests that a common contributing cause of failure during a long-term development project is obsolescence of the technology's anticipated function(s).

In fashioning the case studies—my database—I employ the conceptual tools of the behavioral framework, for they help me ask the most pertinent questions and acquire relevant information.

Perhaps the most important behavioral construct is that of life history. In principle, people and organizations shepherd a successful technology through an idealized sequence of processes: invention, development, manufacture, adoption, and use. Each process usually consists of a set of activities whose performance requirements, when met, enable the technology to advance to the next process. It follows that the life history of a spectacular flop may end during any process. Because recurrent causes of failure tend to be process-specific, life history processes structure the presentation of generalizations.

As already noted, the development of a "revolutionary" technology almost always requires a strenuous effort. The magnitude of this effort—the sum of necessary human, material, and societal resources—is notoriously difficult to predict at the inception of a technology's life history. Thus, in writing about a spectacular flop, we learn about initial estimates of resource needs and how those estimates may have changed. And obviously we find out how, and to what extent, the resource needs were met.

Learning about a technology's functions—anticipated and actual—is essential for constructing generalizations. There are four general kinds of functions.[8]

- *Techno-functions* are utilitarian and are directly involved in manipulating, storing, or transforming matter, energy, or both. Most activities include technologies having techno-functions.

- *Socio-functions* are symbolic and convey social information, such as social power, social role, and social affiliations. Socio-functions, like techno-functions, are ubiquitous in human activities.
- *Ideo-functions* are also symbolic but mainly communicate information about ideology or ideas, such as a religious or political doctrine, and are more common than we might suppose.
- *Emotive functions* evoke fear, nostalgia, joy, and other emotions.

Most technologies—it should be noted—perform multiple functions, often in the same activity.

In contributing to the performance of an activity, a technology must possess relevant performance characteristics: capabilities, competencies, or skills that are exercised (or come into play) in that activity. Performance characteristic is a flexible construct that admits many extensions, as in sensory and financial performance characteristics. A technology's promoters often frame its expected revolutionary impact in terms of performance characteristics that enable new or enhanced functions in important activities. Thus obtaining information on a technology's performance characteristics—anticipated and actual—is required for crafting narratives and laying a foundation for generalizing.

Finally, the behavioral archaeologist is also concerned with identifying the individuals and groups taking part in the activities throughout the technology's life history. These people—originators, promoters, engineers, manufacturers, and users—help establish the performance requirements of activities.

The data requirements for behavioral studies may seem daunting. Fortunately, spectacular flops are often better represented in the historical record than mundane failures. Thus I have been able to draw upon previous research by historians and other scholars to fashion the case studies. These chapters present narratives of a diverse sample of spectacular flops ranging from the world's first self-propelled vehicle to the nuclear-powered bomber. The technologies themselves differ greatly, as do the individuals, companies, and government agencies that participated in their life histories.

In addition to enabling the creation of new generalizations, the narratives themselves are intriguing, for they chart technologies from confident conception to humiliating failure. These life histories are not only compelling from sociocultural and technological perspectives, but they also normalize failure and humanize inventors and engineers.

The names Ferdinand de Lesseps, Nikola Tesla, and R. Buckminster Fuller evoke (respectively) their immensely successful technologies: the alternating-current power system, the geodesic dome, and the Suez Canal. Curiously, these same names seldom evoke failures, though they should—as in de Lesseps' Panama Canal (Chapter 6), Tesla's World System of communication (Chapter 7), and Fuller's Dymaxion houses and car (Chapter 8). Their failures, and those of many other famous creators, are largely missing from our cultural consciousness because the mass media and earlier narratives mainly presented stories about people whose technologies achieved their game-changing goals. Yet the same level of inspiration and perspiration lavished on their successes also went into their failures. Both outcomes result from, and illuminate, processes of technological change.

Because I do not expect the reader to be familiar with every technology discussed below, each chapter includes in narrative form relevant historical information and sketches the technology's basic operating principles. Presented largely in chronological order, the case studies allow the reader to learn about many technologies developed from 1770 to the present.

Chapter 2 introduces the history of steam power, whose origins reach back to natural philosophers of the 17th century. In the late 18th century, the government of France employed Nicolas-Joseph Cugnot on a secret military project to develop a steam-powered dray. Such a vehicle would have eased the great difficulties of hauling heavy artillery from place to place. Using a high-pressure steam engine, Cugnot built two prototypes but the project ended before the second one could be thoroughly tested.

In the 1830s and 1840s, steam-powered railways were a large growth industry (especially in Britain) financed by the sale of shares in companies. Against this background Chapter 3 tells the story of the Samuda brothers' upstart technology, the "atmospheric railway." The Samuda system challenged conventional steam locomotives by employing atmospheric pressure to propel rail cars. The system was collision-proof and was claimed to be cheaper to build and operate. Several very short prototype lines were built in Ireland, England, and France and performed reasonably well. But two longer lines built in England failed: South Devon was never completed and the London & Croydon line was quickly abandoned.

Chapter 4 traces the beginnings of modern electrical technologies, which in the 1830s eventuated in the electric motor. Inventors built their own motors and applied them to various tasks. Only one inventor, Thomas Davenport of Vermont, made a formidable effort to commercialize his motors. Despite receiving a patent and founding a company, his attempt failed. Also in the United States, Charles Page built a reciprocating electric motor—the most powerful of its time—and installed two of them in a locomotive of his own design. To carry out his project, Page obtained a $20,000 grant from the U.S. government. The first trial of Page's electric locomotive, in 1851, was also its last.

As discussed in Chapter 5, Isambard Kingdom Brunel was one of England's most accomplished civil engineers of the 19th century. While building the Great Western Railway in the 1830s, Brunel designed the *Great Western*—the largest steamship of its time and the first to cross the Atlantic entirely on steam power. A second large Brunel steamship, the *Great Britain*, pioneered the use of both an iron hull and screw propeller in the same vessel. In the early 1850s, Brunel envisioned a passenger steamship so large that it could hold enough coal to make a non-stop round-trip to Australia. He convinced a company to back the project and sales of stock supplied the funds. Brunel collaborated with renowned shipbuilder John Scott Russell to construct the 692-foot-long *Great Eastern*, by far the world's largest ship. After only three years of service on the

North Atlantic route, however, the *Great Eastern*'s life as a passenger liner came to an ignominious end.

Chapter 6 is about the canal-building career of Frenchman Ferdinand de Lesseps. He spearheaded—and oversaw construction of—the Suez Canal using private funds raised from stock sales. After the canal's completion in 1869, de Lesseps developed plans for another grand project to connect the Atlantic and Pacific oceans by digging a sea-level canal across Panama. With funds raised from investors, he founded the Compagnie Universelle du Canal Interocéanique de Panama and began the challenging project. After years of slow progress, by 1888 de Lesseps and his company were out of the canal-building business.

Nikola Tesla, a Serbian immigrant who has achieved the status of cult hero, invented the alternating current (AC) motor and multiphase system for distributing electricity. Chapter 7 discusses Tesla's inventions and how they enabled George Westinghouse to assemble an entire AC system that successfully challenged Edison's near monopoly on electric light and power. At the height of his fame, Tesla became convinced that it was possible to build a *wireless* AC system—a World System—for transmitting not only information but also electric power. With funding from J. P. Morgan, he erected a large laboratory on Long Island that included an imposing transmitting tower. Uncharacteristically quiet about the project, Tesla abandoned the station and his dream of supplying inexpensive power and information.

Another independent inventor (and sometime philosopher) who inspired a cult following was R. Buckminster Fuller; several of his projects are treated in Chapter 8. Although Fuller is best known for inventing the geodesic dome, he was a prolific inventor of visionary technologies. His earliest in 1929 was the Dymaxion house, designed to be inexpensive as well as easy to build and maintain, but no company took an interest in manufacturing it. He revised the design after World War II and built a full-scale prototype with U.S. government funding. Despite consumers eager to buy the house, it was not brought to market. Fuller's Dymaxion car, financed with donations from admirers and true believers, was spacious and aerodynamic. Three prototypes were assembled and tested in the early 1930s before the project ended.

In the late 1940s, the U.S. Air Force and Atomic Energy Commission began to invest heavily in the development of a nuclear-powered bomber. As recounted in Chapter 9, such a plane would be able to stay aloft for days (even weeks) without refueling. The attempt to realize this vision was a boon to American corporations, especially General Electric and Pratt & Whitney, which were awarded generous development contracts. During the 1950s the project achieved several milestones, such as showing that a nuclear reactor could power a jet engine on the ground and that a modified B-36 bomber could carry a small operating nuclear reactor in flight. Before showing that it could construct an integrated prototype, however, the project ended in 1961.

Chapter 10 tells the story of the attempt to employ nuclear bombs in large-scale infrastructure projects. In 1953, less than a decade after the attacks on Hiroshima and Nagasaki, President Eisenhower proposed that the United States embark on projects to show that atomic energy could also be used for peaceful purposes. Physicist Edward Teller, who in the meantime had promoted development of the hydrogen bomb, believed that such bombs could become a tool for "geographicical engineering." Teller insisted that bombs could do mega-scale excavations at much lower cost than conventional explosives. Funded by the Atomic Energy Commission, Teller and his associates proposed detailed plans for specific projects, including a harbor in Alaska (Project Chariot) and a new canal across the isthmus of Central America. No project was ever completed.

Another technological spinoff of World War II was the turbojet engine. Chapter 11 recounts how manufacturers around the world attempted to develop turbojet engines that could be installed in land vehicles. Even automakers jumped into the act. The jet-powered car promised simplicity and low maintenance—the engine would have many fewer moving parts—and smoother operation because it directly produced rotary motion. Chrysler made many prototype engines, put them in vehicles, and tested them on the road. What's more, in 1963 it placed fifty demonstration Turbines, a car with an Italian-built body and the latest generation turbine engine, with families across the country. In a follow-up survey, the families reported a fairly high level of satisfaction with the futuristic cars. Despite Chrysler's long-term commitment to the project, the company never mass-produced the handsome vehicle.

International travel has always been a trial, supplying incentives to develop faster conveyances. Chapter 12 tells the story of the Concorde, the supersonic airliner built by a government-financed Anglo-French consortium. The plane was expected to revolutionize overseas travel for wealthy travelers and demonstrate to the world the technological prowess of Britain and France. Beginning in the mid-1950s, engineers in both countries undertook a long-term project involving myriad tests of concepts, components, subsystems, and completed craft. Although the airliner performed superbly, only the national airlines of Britain and France bought the Concorde.

As Chapter 13 details, experiments have been under way since the 1950s to harness the immense energy liberated by the fusion of hydrogen atoms into helium. The conventional approach is to build enormous reactors called tokamaks that operate at more than ten million degrees, but they are not expected to produce commercial power until mid-century (if ever). During the 1980s, however, two chemists—Stanley Pons and Martin Fleischmann—claimed that they had produced "cold" nuclear fusion in a simple, tabletop electrochemical cell. If confirmed, their technology would have been a revolutionary breakthrough promising an endless supply of inexpensive energy. Instead their bold claim generated only scientific controversy. Building on decades of well-funded experiments, a consortium of 35 countries is now spending at least $20 billion to build a giant tokamak in southern France. It is possible that this "hot" fusion reactor may become a flop even more spectacular than cold fusion.

NOTES

[1] Schiffer (2000); Schiffer, Butts, and Grimm (1994).
[2] Kirsch (2000); Mom (2004).
[3] Schiffer (2008a, 2013a:86–88).
[4] These kinds of generalizations are defined behaviorally (Schiffer 2013a).

[5] Recent statements on behavioral archaeology include Schiffer (2008b, 2010); Schiffer, Riggs, and Reid (2017); Skibo and Schiffer (2008); Skibo, Walker, and Nielsen (1995); Walker and Skibo (2015).

[6] Schiffer (2011); see also Hollenback and Schiffer (2010) and Skibo and Schiffer (2008).

[7] Historians of technology also call for studies of failures—for example, Bauer (2014); Braun (1992a); Gooday (1998).

[8] Schiffer (2011) provides definitions of the functions.

2

SECRET PROJECT

Nicolas–Joseph Cugnot and the First Automobile[1]

After putting animals, wind, and flowing water to work, humans harnessed steam. Most people probably remember, perhaps from old movies, trains being pulled by chugging and hissing and tooting steam locomotives. A few of these heavy-metal monsters are still around, toting tourists on nostalgia railroads. But the pre-World War II locomotive was neither the beginning nor the end of steam power. In 18th century England, towering steam engines accelerated the Industrial Revolution by pumping water from deep mines. A century later steam energized many factories, steam railroads crisscrossed the continents, and steamships crowded ports around the world. And to take a recent example, nuclear reactors employ steam-driven turbo-electric generators.

For three centuries in the West, steam power has played a pivotal role in industrial technologies, but it was not successful in every application. Perhaps the most spectacular flop was the automobile, the first attempt to apply steam power in transportation. Two decades before the French Revolution, military engineer Nicolas-Joseph Cugnot developed a small steam engine and mated it to a *fardier*, a dray for hauling heavy artillery pieces. This ungainly vehicle—and especially its unique steam engine—might have revolutionized land transportation, but it had no immediate progeny. Let us examine the historical context of Cugnot's creation and learn more about the history of steam technology.

✳ ✳ ✳ ✳ ✳

The ancient Greeks are sometimes credited with inventing a steam engine, but this claim is misleading. Heron of Alexandria made a device called an aeolipile, which merely illustrated the expansive power of steam (Figure 2.1). It was a metal ball held by a pipe on each side that allowed the aeolipile to spin on an axis. The pipes also conveyed steam to the ball from a boiler below. Steam exited through two small pipes bent at right angles, causing the ball to rotate. It was an ingenious, jet-powered philosophical device, but it was not a steam *engine*, for the aeolipile powered no machine. Curiously, the first steam engines would exploit the contraction—not expansion—of steam.

The earliest steam engines resulted from the research of natural philosophers, whom we would refer to as physical scientists. Renaissance natural philosophers grappled with the notion of a vacuum and eventually dispelled the belief that "nature abhors a vacuum." Beginning with Galileo these men showed that air has weight, which—as we know today—exerts an omnipresent pressure of 14.7 pounds per square inch. Thus when an evacuated vessel is opened, the "whoosh" we hear is the inward rush of air caused by atmospheric pressure.

Perhaps the most colorful figure of 17th century natural philosophy was Otto von Guericke, the *Burghermeister* (mayor) of Magdeburg in Prussia. Trained as an engineer, he is best known for inventing the air pump around 1650. The air pump could produce a partial vacuum that enabled Guericke to exploit atmospheric pressure to perform dramatic public demonstrations. His most famous one took place at Regensburg in 1654 before members of Ferdinand III's Imperial Diet. Wishing to impress these officials with the weight of air, he joined—by means of a leather gasket—the edges of two copper hemispheres, making a sphere nearly two feet in diameter. After pumping air from the sphere, Guericke harnessed a team of eight horses to each hemisphere. Such is the force of atmospheric pressure that the horses, struggling and straining, were only some-

Figure 2.1. Heron's jet-powered philosophical device.
Source: Thurston (1891:Frontispiece).

times able to pull the hemispheres apart. Yet Guericke could easily part the hemispheres by opening a stopcock.[2] After Guericke no person could credibly deny the power of atmospheric pressure, but could machines be created to exploit that power?

Later in the 17th century, French physician Denis Papin turned his attention to natural philosophy and the invention of mechanical devices.[3] One of his earliest inventions, made when he was associated with the Royal Society of London for Improving Natural Knowledge, was the "digester," the first pressure cooker. It could dissolve bone and even had a safety valve. Later, collaborating in Paris with natural philosopher Christian Huygens who was famous for his work in optics, Papin explored the use of gunpowder for creating a vacuum in a cylinder. After concluding that gunpowder produced an imperfect vacuum, Papin—having become professor of mathematics at the University of Marburg in Hesse—turned to steam. In his 1690 report, Papin described a device that used the condensation of steam to move a piston in a small metal cylinder.[4] Although Papin believed that his invention could be employed to pump water and propel ships, he apparently made no engines, for his invention was merely a proof of principle.

One of the first people inspired by Papin's findings and their implications was Englishman Thomas Savery, a military engineer.[5] Savery wanted to make an engine that could pump water from deep, flood-prone mines. Self-financing his project, he came up with a creative design. Savery's engine did not use a piston in a cylinder; instead an enclosed boiler—lacking a safety valve—supplied steam alternately to two vessels (Figure 2.2). Condensation of steam in the vessels produced the pump's suction. With valves that he opened and closed by hand, Savery showed that his engine could raise water. In 1698 Savery received a very general English patent for his invention, which Parliament later extended to 1733. To publicize the invention, he demonstrated a small model to King William and members of the Royal Philosophical Society of London.[6]

In a 1702 book illustrated with playful cherubs, Savery detailed the working of his engine, proclaiming it to be the "miner's friend." He also suggested other potential uses that required the raising of water such as "water-mills, palaces, . . . draining fens, and supply-

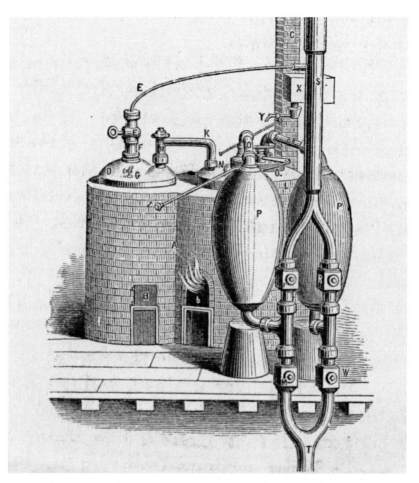

Figure 2.2. Thomas Savery's condensing steam engine (1698).
Source: Thurston (1891:37).

ing houses with water in general."[7] And he claimed it could put out fires.

Carefully managed, Savery's pump worked, but mine owners were not impressed, for it had significant performance shortcomings. The boiler could explode, it might ignite gas in a coal mine, a skilled operator was required, and—most devastating—it couldn't raise water very high. To drain water from a 300-foot-deep mine, for example, would have required five pumps placed at sixty-foot vertical intervals. A handful of one-off applications of Savery's pump for light work were more successful, but no great demand for them emerged.

Savery's engine was not a commercial success but it did show how the condensation of steam in a closed vessel could create enough vacuum to do work by means of atmospheric pressure. This principle would animate the next generation of steam engines ushered in by Thomas Newcomen, an ironmonger, and John Calley, a glazier.[8]

Although neither man had studied natural philosophy, Newcomen and Calley consulted with others and applied their mechanical expertise to make an engine more powerful, versatile, and safer than Savery's. Their first working engine was installed near Dudley Castle around 1712. An engraving shows this machine standing more than thirty feet tall, anchored on two sides by robust brick walls (Figure 2.3). The heart of the engine was a tall brass cylinder containing a brass piston. The piston was connected to one end of a long wooden beam; the beam's other end was attached to the pump. Enclosed in layers of brickwork, a substantial copper boiler was heated by a firebox below and fed steam to the cylinder. The steam's slight pressure and the upward pull of the beam (its other end pulled down by gravity) raised the piston. To create condensation, cold water was sprayed into the cylinder. The sudden vacuum caused the piston to move smartly downward, creating the power stroke that rocked the beam and worked the pump. A Newcomen engine made about 10–25 strokes per minute.

First-time observers of the Newcomen engine at work must have been entranced. Not only did the engine pump water, but it did so automatically through the use of valves and mechanical linkages driven by the engine's own motion. The engine worked powerfully and with a safety valve on the boiler. Instead of seeking his own patent, Newcomen framed his engine as an improvement on Savery's. By the way, steam engines at this time were called either "atmospheric" or "fire" engines, for they depended on atmospheric pressure and on the heat generated by burning wood or coal.

While Newcomen engines were beginning their conquest of flooded mines, Jacob Leupold—a German natural philosopher, instrument maker, and engineer—published a series of ten handsomely illustrated books that featured his mechanical inventions,

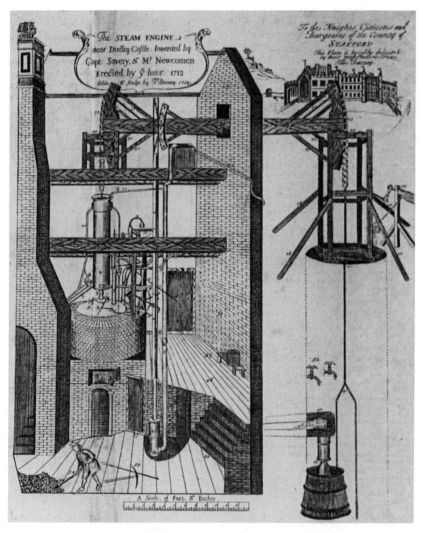

Figure 2.3. Thomas Newcomen's condensing steam engine (1712).
Source: Galloway (1881:Plate I).

some of them quite fanciful (as in perpetual motion machines) yet many more feasible. In his 1724 book *Theatri Machinarum Hydraulicarum* (*Theater of Hydraulic Machines*), he offered a radical design for a steam engine. Instead of using a vacuum to suck the piston downward in the cylinder, his engine would use high-pressure steam to push a piston upward. His design included an ingenious rotating valve that alternated the steam between two cylinders (Figure 2.4).

An engine that exploited expanding steam could exert great pressure and have many fewer parts. However, inventors may have been reluctant to risk boiler and cylinder explosions. There is no evidence that Leupold himself actually built a high-pressure steam engine, but in the ensuing decades his elegant design would not go unnoticed.

In the meantime the Newcomen engine promised to satisfy the needs of coal companies. But was the Newcomen engine really a cost-effective replacement for horses and men for pumping water? Famed natural philosopher John Theophilus Desaguliers reported that the annual expense of an engine at Griff (near Coventry) was £150, compared with £900 for the fifty horses it replaced.[9] A 1739 account by Bernard Forest de Bélidor, a French hydraulic engineer, tells of the heroic work done by a Newcomen engine in France.

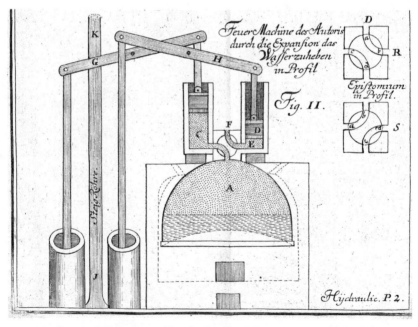

Figure 2.4. Jacob Leupold's design for an expansion steam engine. Source: Leupold (1724:Table XLIII, Figure II). Courtesy of Dibner Library, Smithsonian Institution.

"Previous to the erection of this engine, fifty horses and twenty men, working day and night, had been required to raise the water from the mine, whereas the engine with a single attendant, in forty-eight hours working, cleared the colliery of water for a whole week." De Bélidor went on to pronounce it "the most marvelous of all machines."[10] The engine's great economy depended on the fact that fuel expenses in a coalmine were negligible.

Throughout the British Isles and other countries, many coalmines were outfitted with Newcomen engines: rising water was kept at bay and abandoned mines were worked again.[11] After the Savery patent—which had covered the Newcomen engine—expired in 1733, other entrepreneurs began building and installing steam engines. Between 1734 and 1780, more than six hundred were constructed in Britain alone.[12]

Clearly, coalmines were low-hanging fruit for makers of steam engines to pluck. Extending the engines to other applications was a different matter because in some regions coal was very pricey—and the Newcomen engine had a voracious appetite for coal. This problem attracted the attention of James Watt, the University of Glasgow's mathematical instrument maker. Asked to repair a small model of a Newcomen engine during the winter of 1763–1764, Watt (who was well versed in natural philosophy) became intrigued by the engine's wasteful use of heat.[13]

To learn more about the properties of steam, Watt embarked on a series of experiments. He learned, for example, that steam occupied a volume around two thousand times greater than the water that produced it. More important, he also identified a major cause of the Newcomen machine's excessive fuel consumption: much steam was wasted through the alternate heating and cooling of the cylinder. Accordingly, he concluded that the engine's cylinder should always remain hot—as hot as the entering steam.

His solution to this problem was elegant in theory: he would use a separate vessel to condense the steam. He would also place a steam jacket around the cylinder, which allowed it to remain hot. The condenser would be immersed in cold water and remain cold.

But in practice this solution created a host of subsidiary problems, such as the design of new valves and pipes to transfer the steam among the components. In the course of experiments with models, Watt had to make many modifications. In addition, the new engine design placed greater demands on component tolerances, especially the cylinder's bore.

Not a man of wealth, Watt incurred a large debt in conducting his experiments. He found a partner, physician-turned-entrepreneur John Roebuck, who assumed the debt and acquired a two-thirds share of the invention. A patent was issued on January 5, 1769, for "a new method for lessening the consumption of steam and fuel in fire engines."[14] When Roebuck ran into financial difficulties and could no longer support the project, Watt found another partner—Matthew Boulton, a manufacturer of metal products in Soho near Birmingham, who bought out Roebuck.

Watt moved to Soho and, after years of further work and re-newal of the patent for 25 years, Boulton's factory produced its first Watt steam engines in 1776. More economical to operate than Newcomen engines, Boulton–Watt engines enjoyed brisk adoptions for many long-envisioned applications—including draining mines where coal was costly, driving machinery in factories, operating mills, draining fens, and pumping water for municipal systems. By 1800 around 450 Boulton–Watt engines had been installed in Britain.[15] However, Boulton–Watt engines were huge and heavy, disadvantages where space was at a premium.

For the well-traveled person of the late 18th century, steam engines were a noticeable feature of rural landscapes. Surely some people seeing these machines at work would have pondered further possibilities for applying their immense power. No doubt ideas for steam-powered transportation (coaches, wagons, railroads, boats, and ships) were sufficiently obvious to have occurred to many people. Usually, in a technology's early history people envision new uses, but creating prototypes and sustaining development projects requires resources. Ideas are cheap; development projects often are not. Papin was able to carry out his experiments with some support from philosophical societies and the University of Marburg; Savery was self-financed; Newcomen and Calley were also self-financed at first, relying on wealth created by their businesses; Watt, of limited means, found partners—first Roebuck, then Boulton—to support his work. Even before the modern era, inventors had some ways to finance their projects.

Beginning in the late 18th century, people with promising ideas had greater opportunities to acquire funds because the expansion of industry and commerce was creating new sources of wealth, forms of organization, and kinds of expertise. Such wealth allowed people like Boulton to invest in promising inventions. Also, entrepreneurs could take advantage of corporate organization and—along with more traditional partnerships and companies—employ forms of financing coming into wider use, such as the issuance of stocks and bonds and the sale and licensing of patents. In addition, manufacture of the steam engine and other industrial products had contributed to the growth of expertise in (and equipment for) precision working of metals on a large scale. And throughout history and even prehistory, governments had supported technology projects from pyramids to city walls, and they would continue to do so.

A great variety of technological, economic, and human resources was now available to develop promising ideas, such as steam-powered transportation. Between 1770 and 1830 inventors and entrepreneurs cobbled together resources, launched projects, and created steamboats, steamships, railroads, and automobiles. In many projects an engine with a high power-to-weight ratio was required. In these applications Boulton–Watt engines were generally eschewed in favor of high-pressure engines similar to the one designed by Leupold. The first of the new technologies to reach the prototype stage was the French steam-powered automobile. This was surprising because, compared with Britain, industrialization in France was moving slowly. Nonetheless, the world's first motor vehicle was built in Paris and financed by the French government.

In 1769, the same year that Watt's steam engine was patented, Nicolas-Joseph Cugnot—a French military engineer—published his third book on warfare, a treatise on fortifications.[16] Having served in the Seven Years War (1754–1763), Cugnot had acquired much practical experience in land combat and developed ideas for improving military equipment and transportation. As an instructor at the Paris Arsenal, Cugnot began designing a *fardier à vapeur* (steam-powered dray).[17]

The idea of a steam *fardier* had already been under consideration by French officials. Lieutenant General Jean-Baptiste Vaquette de Gribeauval was a reformer trying to standardize artillery parts to make them interchangeable; he was also interested in finding better ways to transport the largest cannons across the countryside. *Fardiers,* pulled by a team of oxen sometimes helped by horses, were used for this purpose. Gribeauval chose Cugnot—a man highly regarded and already at work on the project—to build a small model. This pioneering attempt would require that a steam engine's reciprocating motion be translated into rotary motion, a feat that even Watt had not yet accomplished. But Cugnot had to devise such a mechanism if his vehicle were to run.

Sometime during his many years of military service, Cugnot had come across Leupold's 1725 design for a compact, high-pressure steam engine with two cylinders. Using an engine along these lines probably seemed preferable to downsizing a Newcomen-style engine, which might have failed to produce enough power to move a multi-ton vehicle. In addition, the construction of a Leupold engine might be simpler owing to its fewer components.

At the army's expense, Cugnot developed the model quickly and it was ready the following year for a trial before Gribeauval and France's Foreign Minister Choiseul. With four people aboard it moved at about four kilometers per hour—equal to a leisurely walk. However, the model couldn't maintain steam for more than 12–15 minutes. When the vehicle periodically came to a stop for lack of steam, water had to be added and the fire renewed. Steam could not be maintained while the *fardier* was in motion because the firebox-boiler was cantilevered past the front end, its opening unreachable. The periodic stoppage and long wait for steam to rise again was a severe performance defect. Even so, the model's shortcomings were not judged fatal and Cugnot secured authorization to build a full-scale *fardier.*

The new steam dray was built at the Paris Arsenal according to Cugnot's design but under the direction of Michel Brézin, a noted locksmith. The components were made at the arsenal except for the bronze cylinders and pistons, which were cast at a foundry in Strasbourg. Each cylinder was fourteen inches tall and twelve inches in diameter.[18] With three wooden wheels, the engine fixed above the single front wheel, and the firebox-boiler suspended out front, it had a most distinctive appearance (Figure 2.5). The vehicle was completed sometime in 1770.

The first person to follow Leupold's design for a high-pressure engine, Cugnot employed steam expansively. The two cylinders were inverted, with the pistons pushing downward alternately on complicated linkages that drove the front wheel from both sides (Figure 2.6.). Each stroke of a piston rotated the wheel one-quarter turn in forward or reverse. To stop the *fardier*, the driver operated a foot lever that pressed a wooden shoe against the rim of the front wheel.[19]

The innovative rack-and-pinion steering was operated with a handlebar-like control; the gearing permitted it to compensate for the weight of the engine and boiler on the front wheel. Like its predecessor, the *fardier* reportedly crept along at about four kilometers per hour. According to one account, the new version did not suffer a steam deficit as severe as the earlier model because it had a much

Figure 2.5. Nicolas-Joseph Cugnot's *fardier*.
Source: Figuier (1891:Figure 123).

Figure 2.6. Driving mechanism of Cugnot's *fardier.*
Source: author's photograph.

larger firebox-boiler. Unfortunately, the second vehicle did not undergo final testing and there are no reliable reports of its performance characteristics. Clearly, major refinements would have been required before the steam *fardier* could have entered production and been sent into battle.

Perhaps later-generation prototypes might have overcome the vehicle's performance shortcomings, but this was not to be. Gribeauval fell out of favor with the government and his reform efforts stalled; Choiseul was also gone. Apparently, no one in power saw an urgent need for a self-powered *fardier*—after all, France was not at war. Without a government patron willing to champion the project and no compelling military need, it died. Cugnot's creation, which had set so many technological precedents, was stored in the Paris Arsenal and forgotten for decades.[20]

Because the steam *fardier* was a military project, few people knew about it. The first British and American accounts of the *fardier* did not appear until the middle of the nineteenth century.[21] That Cugnot's project was largely secret helps to explain why no one continued where he left off.[22] Had other inventors and entrepreneurs known about the success of a safe and compact high-pressure steam engine, the development of steamboats and railroads might have been accelerated. But they had no such knowledge and Cugnot's engineering feats had to be reinvented in later decades.[23]

＊ ＊ ＊ ＊ ＊

I first became aware of Cugnot and his steam *fardier* when my 1994 book *Taking Charge: The Electric Automobile in America* won the Cugnot Award of Distinction from the Society of Automotive Historians. I was deeply honored and of course felt compelled to learn a little about the award's namesake. I didn't think much about early automobiles over the next two decades or so, but then Cugnot came into my life again—another pleasant surprise.

On a trip to Paris in 2012, I visited the Conservatoire National des Arts et Métiers (National Conservatory of Arts and Crafts). In addition to being an elite degree-granting institution, the conservatory has a fabulous museum of technology that most tourists miss. It opened in 1802 in the deserted Priory of Saint-Martin-des-Champs. As I wandered nearly alone through its galleries, marveling at the technological gems that had been preserved, I came across Cugnot's *fardier*. To say the least, I was thrilled; I had no idea the vehicle still existed. Now that I'm writing about it, I regret not taking copious notes and many more pictures. In any event, the world's first automobile is there for all to see, rescued from the Paris Arsenal more than two centuries ago. Visitors to the museum must wonder how well it ran—indeed, if it ran at all. Alain Cerf recently took up that question in earnest.

An engineer and owner of Polypack, a company "that designs and manufactures automatic packaging machinery," Alain Cerf is an automobile enthusiast.[24] To house his collection of historical vehicles that showcase innovative engineering, he built the Tampa Bay Automobile Museum, which opened in 2005. I suspect the dream of every automobile collector is to own Cugnot's *fardier*. That being impossible, the next best thing is to build a replica, and this is precisely what Cerf did, aiming "to construct an exact working replica of the *Fardier*."[25] The survival of Cugnot's creation has made it possible to build and test a fairly faithful replica in the absence of the original plans. Fortunately, Cerf did discover—in the library of one Monsieur Vallière—a detailed parts description of the vehicle written about 1770.[26]

Cerf's first move was to visit the Conservatoire National des Arts et Métiers. In addition to documenting the vehicle itself in intimate detail, he and his team copied materials from the conservatory's archives. His second move was a visit to the Deutsche Bahn Museum in Nürnberg where a 1930s nonworking replica of the *fardier* resides. Remarkably, Cerf managed to convince museum officials to allow his team to install a steam engine in the replica—originally it ran on compressed air—and to display it briefly in Florida. The German replica with its new engine is now back in the Deutsche Bahn Museum. These moves furnished much of the information Cerf and his team needed to build an operational Cugnot *fardier* and also provided new insights into the vehicle's original design.

Before construction could begin, Cerf had to study the conservatory's *fardier* closely to pinpoint any post-Cugnot modifications. In fact, he found more than a handful of troublesome anomalies—including missing components such as a boiler valve, the mechanism to distribute steam to the cylinders, and many bronze parts. He attributed the anomalies to the vehicle's transfer from the arsenal to the conservatory and to changes made prior to an exhibition in 1907. There is a strong suspicion that the museum furnished Cugnot's *fardier* with some facsimile components, such as the pistons, which were incorrect. And the firebox-boiler could not be the original, for it was too small and incapable of supplying sufficient steam—an interesting puzzle.

Cerf devised a plausible scenario to explain the undersized firebox-boiler that accords with an old story about the world's first automobile accident. Supposedly, after Cugnot's project ended, the full-scale *fardier* was accidently driven into a wall at the Paris Arsenal (see Figure 2.5).[27] If this story is true, then the original firebox-boiler might have been damaged beyond repair. Cerf suggests that it was replaced by the undersized firebox-boiler from the 1769 model, which it carries to this day. Informed by modern judgment and imagination, the Cerf team resolved the anomalies and solved the puzzles, drawing upon Vallière's parts list and on the engineering knowledge and technology of the late 18th century to build a Cugnot *fardier* with an ample firebox-boiler.

After about a year's work, the Tampa Bay Auto Museum's replica was completed in late 2010.[28] It underwent testing in early 2011 and was declared fit for the road (Figure 2.7). Since then it has been exhibited at fairs and automobile shows in America and Europe, such as the Avignon Motor Festival and the Essen Motor Show. Drivers wear 18th century costumes including a tricorn hat. Many YouTube videos show the wood-burning vehicle in operation, a cannon barrel slung underneath, puffing and clanking and crunching its way across the pavement at a walking pace.[29]

According to Cerf, this working replica promotes the belief that Cugnot was a brilliant engineer and that France, ostensibly a backwater in 18th century mechanical engineering, was in fact on the

Figure 2.7. Alain Cerf's replica of Cugnot's *fardier*.
Source: courtesy of the Tampa Bay Automobile Museum.

cutting edge of a new and potentially significant technology. These messages are no doubt readily received by Americans, Britons, and Germans who attend the fairs and shows where the vehicle is demonstrated. And in France the functioning *fardier* boosts national pride in the accomplishments of a homegrown engineer-inventor. None of this is surprising, given that the man who brought the Cugnot *fardier* to life happens to be French-born Alain Cerf.

NOTES

[1] I thank Roger White, curator of transportation at the National Museum of American History, Smithsonian Institution, for providing several sources on Cugnot's *fardier*.

[2] von Guericke (1994:160).

[3] On Papin and his engines, see Galloway (1826:13–15) and Galloway (1881:26–51, 72–77); see also Thurston (1887).

[4] Galloway (1881:47–51).

[5] On Savery and his engines, see Galloway (1881:57–71) and Savery (1829).

[6] Savery (1699).

[7] Savery (1829:31).

[8] On the Newcomen engine, see Galloway (1826:15–17, 19) and Galloway (1881:78–90).

[9] Galloway (1881:105–106).

[10] de Bélidor, quoted in Galloway (1881:116–117).

[11] Galloway (1881:91–120) discusses the early commercialization of the Newcomen engine.

[12] Friedel (2007:197).

[13] On Watt's modified steam engine, see Galloway (1881:126–149).

[14] Galloway (1881:148).

[15] Tann (1981:6).

[16] Cugnot (1769).

[17] Unless otherwise noted, information on Cugnot and his two vehicles comes from Cerf (2010), Conservatoire National des Arts et Métiers (1956), and Morin (1851). Fletcher (1891:17–19) describes an English effort that was roughly contemporary to that of Cugnot.

[18] Cerf (2010:20)

[19] Cerf (2010:61–62). The *fardier's* mechanical components are illustrated

and their operations described in a book targeted at youth (Jacomy and Martin 1992).

[20] Cerf (2010:63) lists the technological precedents but there is no historical continuity between them and their reinvention decades later.

[21] Cowper (1853); Howe (1844). It is possible that Morin (1851) was the first published notice of Cugnot's work in France. That source reproduces two important letters (L.-N. Rolland, 24 January 1801; Gribeauval, 23 April 1770) whose contents appear to have been the basis of many later accounts.

[22] Cerf (2010:20) makes this point.

[23] Cerf (2010:81–86) describes one possible exception: Charles Dallery.

[24] http://www.tbauto.org/about.htm, accessed 23 September 2016.

[25] Cerf (2010:6).

[26] Published in English by Cerf (2010:32–34).

[27] In another version of the accident story, the *fardier* fell over while turning a corner (Cowper 1853:301).

[28] It was built by David Falvey, Andy Kinworthy, and Alain Montpied (Cerf 2010:3).

[29] https://www.youtube.com/watch?v=lYWnxUxZlmw, accessed 26 September 2016; https://www.youtube.com/watch?v=C3p55J-VA5k, accessed 26 September 2016; https://www.youtube.com/watch?v=Xs EbeCrPcA8, accessed 26 September 2016.

3

TOO LITTLE, TOO LATE

The Atmospheric Railway

During the 1840s, when British steam-powered railways were laying down tracks and building rolling stock at an unprecedented pace, along came a competitive system. The "atmospheric" railway was a radical departure from steam-powered railways because it depended on atmospheric pressure to propel a train. Its promoters, who included famed civil engineer Isambard Kingdom Brunel, claimed that it would revolutionize rail travel. Let us begin by situating the atmospheric railway in the decades-long development of early British railways.

✳ ✳ ✳ ✳ ✳

During the last decades of the 18th century, several enterprising companies established railways in England. They took advantage of the principle that a wagon or carriage running on iron rails or plates encounters less friction than one running on a bumpy, rutted dirt road. Instead of dragging one wagon over treacherous roads, the same team of horses—the motive power in the first railways— could haul ten wagons on an iron road. The Surrey Iron Railway, completed in 1802 in what is now suburban South London, was one of the first public railways. Shippers used the nine-mile line to move goods and raw materials, including coal—the fuel for homes and many factories.[1]

Steam-powered locomotives promised a way to move longer and heavier trains, perhaps more quickly, on iron roads. Stationary steam engines had already proved their worth in mines and factories and the idea of putting a steam engine on a carriage or wagon was just a short leap of imagination. Adapting a steam engine for driving a wagon required much creative engineering, but there was no shortage of effective designs, especially those exploiting high-pressure steam. A second car—called a tender—would have to follow the locomotive, carrying a supply of coke (an expensive fuel, derived from coal) and water to feed the boiler that supplied steam to the engine.

During the teens several collieries (coalmines) experimented with steam-powered railways. The manager of the Middleton Colliery, John Blenkinsop, commissioned machine makers Murray and Wood to build a locomotive.[2] Completed in 1812, the *Salamanca* hauled coal wagons from the mine to the nearby city of Leeds. The Middleton Colliery Railway was the first successful use of a steam locomotive for commercial purposes. Other collieries soon followed.

George Stephenson was a man whose humble origins and limited education did not bar him from high attainments in the nascent railway industry.[3] Having worked in several collieries, Stephenson learned first hand about steam engines and railways. As an employee of the Killingworth Colliery, Stephenson designed and built a locomotive that began service in 1814. Weighing six tons, Stephenson's *Blücher* could pull wagons laden with as much as thirty tons of coal at four miles per hour, even up a gentle rise. This railway was so successful that Stephenson was hired to build locomotives and design a railway for the Hetton Colliery, which began operating in 1822.[4]

Stephenson was also named engineer in charge of constructing the Stockton & Darlington Railway in northeast England.[5] Like all early railways, it was a speculative enterprise whose construction was funded by the sale of shares. One of the largest shareholders was Edward Pease, who became the company's director. After organizing the company, its promoters had to secure an act of Parliament before work could begin. Parliamentary debates gave people whose land was crossed by the railway an opportunity to express their opinions. If powerful landowners raised objections, the route might be modified—as resulted from debates over the Stockton & Darlington Railway. After several tries the act was at last approved in 1821. Every railway in Great Britain had to undergo this political process, which sometimes lasted many years.

Partnering with Pease, Stephenson established Robert Stephenson & Company in Newcastle to build locomotives. The firm prospered in subsequent decades and eventually employed many hundreds of men. Built by Stephenson & Company, the Stockton & Darlington locomotives were highly capable. With Stephenson himself at the controls, the first locomotive (*Locomotion*) hauled an eighty-ton load of mainly coal cars for nine miles, reaching a top speed of 24 miles per hour. On another early test run, *Locomotion* pulled a customized passenger car—a railway first.

The Stockton & Darlington ran for 25 miles on wrought iron rails, which were much less brittle than cast iron. The rails were 4 feet, 8.5 inches apart, which became the most widely adopted standard gauge. When it opened in 1825, the Stockton & Darlington was the first steam-powered *public* railway in the world. And it was profitable.

Although the Stockton & Darlington was an important pioneering line, the Liverpool & Manchester Railway received more attention and became the model for *intercity* railways.[6] Liverpool was a very busy port on England's northwest coast; Manchester, inland, was a major manufacturing city. Connected by canals, the cities supported a steady stream of cargo: raw materials from Liverpool went east, finished goods from Manchester went west. Perhaps, promoters in both cities believed, a railway line would offer less expensive service than canals. Accordingly, a company was organized and shares were sold to 308 investors living in Liverpool, Manchester, and even London. Parliament approved the railway in 1826 and Stephenson and his son Robert, a university-trained engineer, soon became the project engineers.

By the time George Stephenson assumed this position, he had concluded that railways could only succeed if their lines were built as level as possible. To achieve this ideal, the 35-mile-long Liverpool & Manchester line required 64 bridges and viaducts as well as two long tunnels, one of which ran more than a mile under Liverpool to the docks. Due to the many construction projects required, and because level lines in hilly country were expensive to build, the door was open to competing technologies that promised lower "first costs."

The Liverpool & Manchester line consisted of two parallel tracks that supported travel in both directions. Robert designed new locomotives, which had a multi-tube boiler of greater efficiency than earlier models. Of the many locomotives that were needed, the first one—and the one most closely identified with the railway—was named *Rocket* (Figure 3.1).

To ensure that their railway had the best available equipment, the company established a contest in which the *Rocket* had to com-

Figure 3.1. George Stephenson's rocket (ca. 1829).
Source: Findlay (1889:XVII).

SPECTACULAR FLOPS

pete against other locomotives for a prize of £500. Four locomotives were eligible for the trials, which began on October 6, 1829, in Rainhill, on a level stretch of track near Liverpool. The contest was a grand Georgian spectacle, attracting thousands of people, including women of "beauty and fashion" who were seated in stands.[7] The reliable *Rocket*, hauling thirteen tons, completed the seventy miles of trials averaging a conservative twelve miles per hour, but it once reached thirty miles per hour. Only the *Rocket* was still running at the end of the trial and won the prize. Robert Stephenson & Company was awarded a contract to build locomotives for the Liverpool & Manchester.

The ceremonial opening of the Liverpool & Manchester line, on September 15, 1830, was attended by many dignitaries who rode in the train carriages—including the Duke of Wellington, then the British prime minister. The festive occasion was marred by the death of William Huskisson, a local member of Parliament, who fell under the wheels of a passing locomotive. Fatal accidents, including collisions between trains, would bedevil railway travel for decades.[8]

From the very beginning, the line catered to passenger travel by scheduling trains of several classes pulled by *Rocket*-style locomotives. First class was passenger-only coaches, fully enclosed; in second class were passenger coaches open to the elements; and third class was an assortment of freight cars. Many trains ran daily and the company set up a signaling system to prevent trains from crashing into others stopped on the same track. Stationed at intervals along the line, men used hand signals to indicate whether a train could proceed safely.

The Liverpool & Manchester Railway was immediately profitable (it paid shareholders an annual dividend of 9.5 percent), revealing a hefty demand for intercity railways that could carry both cargo and passengers. By the 1840s Parliament had received more than six hundred petitions for similar railways. In mid-decade the enthusiasm reached a fever pitch—a "railway mania" it was called—sending stock prices for companies soaring. Although the bubble soon burst, this didn't affect railway expansion. In 1846 there were 3,036 miles of railways in Britain; by the end of the decade there were 6,031 miles, connecting major cities and many minor ones.[9] The Stephensons, widely acclaimed for their engineering prowess, designed several dozen railways, spreading far and wide the 4-foot, 8.5-inch gauge.

During the late 1830s, Isambard Kingdom Brunel was building the Great Western Railway, which was intended to connect Bristol and London. Brunel had very definite ideas about railway design that put him at odds with the Stephensons, for the Great Western employed a seven-foot gauge. Not only would the wider gauge give a smoother ride, insisted Brunel, but it also allowed the locomotive and cars to be slung low. The lower center of gravity made trains safer, less likely to jump the tracks. Unfortunately, trains built for one gauge could not run on the other, thus creating a travel barrier that required the transfer of passengers and cargo between trains. (Decades later a third track was added to wide-gauge lines, allowing trains designed for the now "standard" Stephenson gauge to run on them.)

Despite the Stephensons' growing dominance in the railway industry, Brunel sought new business—especially in the south and southwest of England, where he could lay down wide-gauge tracks and connect with the Great Western. He also explored railway opportunities farther north, where he was competing directly with the Stephensons. Their competition, which ramped up in the early 1840s, became known as the battle of the gauges. It was during this decade that the atmospheric railway had its brief heyday.

✳ ✳ ✳ ✳ ✳

Brunel did not invent the atmospheric railway. In fact, he was the last British engineer to employ it. The idea of propelling a train by air pressure had been around for many decades.[10] The atmospheric railways built in the 1840s followed the basic plan set forth by Henry Pinkus, an American living in London.

Imagine an iron pipe, say 10–15 inches in diameter, set on the ties between the rails. A narrow slit, perhaps two inches wide, runs along the top of the pipe for its entire length, covered by a longitudinal valve, a continuous strip of leather fixed to one side of the slit. Inside the pipe is a tight-fitting piston. A vertical arm bolted to the piston passes upward through the leather valve and attaches to the train's first car, called the "piston carriage." When a steam-driven pump removes air from the pipe in front of the piston, creating a partial vacuum, atmospheric pressure pushes the piston from behind and carries the train forward. The key to making this technology work is the valve that opens momentarily when the arm arrives and closes immediately after it goes by. Although he had obtained a patent for his invention, Pinkus was unable to form a company to build his "pneumatic" railway.[11]

Engineers Samuel Clegg and Jacob Samuda were associates of Pinkus. Samuda and his brother Joseph, sons of a prosperous merchant, owned a successful company in London that built marine steam engines. Clegg and the Samudas took Pinkus's basic idea, made important modifications—especially to the valve mechanism—and received a patent.[12] Their next step was to build a short test line. In 1838 they obtained permission to modify an unused portion of track at Wormwood Scrubs belonging to the Birmingham, Bristol & Thames Junction Railway Company.[13] With their own resources, during a two-year period Clegg and the Samudas built an atmospheric railway one-half mile long. The test track had a slight slope, rising about one foot per 120 feet traveled.

Nine inches in diameter, the iron pipe was made of segments having deep socket joints packed firmly with materials—perhaps with rope and tar—that maintained the vacuum. A stationary steam engine of sixteen horsepower made in the Samuda factory powered the vacuum pump.

To visualize the design of their valve, consider what would happen if it consisted only of leather. In this case the vacuum would suck it into the pipe, rendering it useless. To keep the flexible leather where it belonged, Clegg and the Samudas sandwiched it between two plates of iron, in segments about eight inches long that ran along the pipe's entire length. One piece of iron was attached to the underside of the leather, which when the valve was closed completed the circle of the pipe, helping to maintain the vacuum. The second iron strip, attached to the top of the leather, was wider than the slit and rested on a flange; it served to keep the closed valve in place (Figure 3.2).

Since leather by itself is an imperfect seal, the unattached side on the flange was likely to leak. To handle this problem, Clegg and the

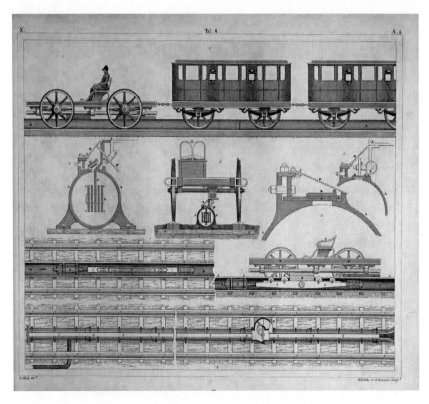

Figure 3.2. Details of the atmospheric railway.
Source: Heck (1851:429).

Samudas coated the flange with a mixture of molten beeswax and tallow, making an airtight seal. But the situation was more complicated because the seal had to be broken briefly and then quickly resealed. As the train moved forward, rollers attached behind the piston pushed upward, breaking the seal and opening the valve—which allowed the arm to pass. A roller on the bottom of the piston carriage then pushed the valve downward, reseating it on the flange. To ensure that the valve was sealed again, the sealant was softened immediately. To accomplish this, the inventors slung below the piston carriage a five-foot-long copper tube filled with hot coals that passed just behind the connecting arm and above the valve. Could this ingenious valve mechanism work in practice? And if it worked, could it do so reliably over long periods?

At Wormwood Scrubs, Clegg and the Samudas tested the atmospheric railway during an eight-month period. The trials consisted of 35 runs in twelve days, the train carrying an average of 23 passengers. It traveled at speeds of 18–45 miles per hour—the higher speeds coming late in the trials, suggesting that continuous tinkering was paying off.[14] A full-scale system would, they calculated, require pumping stations at three-mile intervals—consisting of two pairs of steam engines and vacuum pumps (one set for backup), men to operate them, and a substantial building to house all the equipment, including a large boiler.

Joseph Samuda presented the results of the experimental line in an 1841 pamphlet, seeking to entice investors to license the atmospheric system for building real railways.[15] He claimed that first costs and annual operating costs per mile would be less than those of a steam railway. These cost comparisons were shaky given the limited experience on a single track just a half-mile long. Claims for safety were more believable because trains couldn't crash into each other or fall off the track. Thus the atmospheric railway would result in "the large saving of human life and suffering."[16] In addition, the atmospheric railway did not require a level line, since the vacuum could easily—and at moderate speed—pull a train up a slight incline. In theory, then, the atmospheric railway's purported performance advantages heralded a revolution in train travel.

The pamphlet also highlighted the supposed shortcomings of locomotive railways, such as a maximum economical speed of 25 miles per hour and the inefficiency of the locomotive hauling its own great weight plus the tender. Samuda was emphatic that "the evils of the present system are entirely attributable to the use of locomotive power." In a final appeal to potential investors, Samuda argued that only three of seventeen functioning steam railways were yielding a generous return on investment. The atmospheric railway, he promised, would allow investors to realize "a return for their capital which they so richly deserve."[17]

Clegg and the Samudas invited interested parties to witness the running of their little railway, which was conveniently located between London and Wembley. Their invitations were accepted by many railway engineers, cabinet ministers, government officials, and foreign ambassadors; even Prince Albert made an appearance.[18] Visitors could not help but marvel, for the atmospheric system lived up to its advance billing, reaching speeds of 45 miles per hour. An 1840 article in *Railway Times* validated the technology. "[W]e have no hesitation in now bearing our personal testimony to the efficiency of the system."[19] But the article ventured no opinion on the system's economy.

Among the visitors to Wormwood Scrubs were officials from the Dublin & Kingstown Railway, Ireland's first, which would soon be extended from Kingstown to Dalkey—a distance of less than two miles. After witnessing the working of the experimental line, the Irish railway men became enthusiastic about adopting the atmospheric system.[20] Before committing to the project, however, James Pim, Jr., treasurer of the Dublin & Kingstown, requested that a committee of the Board of Trade look into the new technology.

A committee consisting of Frederick Smith of the Royal Engineers and natural philosopher Peter Barlow investigated several issues: (a) whether the atmospheric system could run longer distances; (b) first costs for building a line; (c) operating expenses; and (d) safety. The committee visited the Wormwood Scrubs installation and conducted many experiments. Contradicting the builders, the committee concluded that first costs would actually be greater per mile than a steam railway; if trains ran frequently, however, operating expenses would be less. As to safety, they recommended that a means be found for rapidly stopping the atmospheric train. Noting that there was no obstacle to building longer lines, tellingly they did acknowledge "practical difficulties in regard to junctions, crossings, sidings, and stoppages at road stations, which may make this system of less general application." Despite some reservations, the committee's conclusion—reported in 1842—was encouraging. "[T]he atmospheric principle seems to us well suited for such a line as the projected extension from Kingstown to Dalkey."[21]

On the basis of this largely favorable report, the Dublin & Kingstown obtained a £25,000 loan for the project from the British government; Clegg and the Samudas were the consulting engineers on the 1.75-mile-long Kingstown and Dalkey line. Trials of the single-track system, completed in the summer of 1843, attracted large crowds—including foreign visitors—and even impressed some skeptics in the railway press.[22] The *Dublin Evening Mail* waxed ecstatic over the Irish triumph. "[T]his wonderful and extraordinary development . . . will produce a revolution in the whole system of railway construction and traffic, and in the end be universally adopted."[23] In March 1844 the line opened for business, offering passage from Kingston on the hour and half-hour and from Dalkey at a quarter to and a quarter after the hour.[24] Business was brisk.

The new line had a gentle incline with Dalkey at the higher elevation. On the uphill run, the steam engine and vacuum pump connected to the main pipe through a short pipe and in effect sucked the train uphill. On the downhill run, the train coasted to Kingstown. This meant that the system only needed one pumping station—saving many complications, extra expenses for equipment, and operating costs. The builders had used essentially the same technology as that on the Wormwood Scrubs test, but much was learned about how to keep the system running well. To reduce friction during the piston's travel, the interior of the cast iron pipe was coated throughout with tallow, and leather packing around the piston was periodically replaced. Connections between pipe segments on the roadbed and in the pipe connecting to the vacuum pump were inspected often; the packing, if found leaking air, was repaired.

The longitudinal valve also needed frequent maintenance, an added expense of men and materials. Not only did the leather have to be continuously saturated with grease, but places that leaked air—perhaps the result of inferior leather or poor tanning—had to be replaced, usually in 75-foot segments. Even so, valve repair was routinized, carried out at night, and didn't interfere with the train's daily schedule. The valve sealant—one composition for summer, another for winter—had to be inspected often and replenished or replaced.[25]

Although the valve and pipes leaked anyway, especially at high vacuum, the Kingstown and Dalkey segment—according to contemporary accounts—worked very well and sometimes reached fifty miles per hour, even around the route's fairly tight curves.[26] In the line's first eleven months of operation, an average of 26 round-trips were made daily, carrying a total of 240,000 passengers.[27] A one-way trip took about two minutes. The *Illustrated London News* was effusive in its praise, noting that atmospheric propulsion "afford[s] a cheaper, safer, and a more convenient mode of conveyance, than locomotive railways."[28]

Visitors to the experimental railways at Wormwood Scrubs and Kingston and Dalkey, which included Stephenson and Brunel, saw trains running swiftly (almost silently), belching no smoke or cinders, and carrying passengers in comfort and complete safety. Many people no doubt thought that this strange atmospheric system could be the future of intercity travel and commerce. Excited by the possibilities, dozens of companies proposed plans for railways that might use the system.[29] But only three were actually built: the 1.38-mile St. Germain line in France and two longer lines in England. Famous engineers championed and built the long lines: William Cubitt on the London & Croydon Railway and Isambard Kingdom Brunel on the South Devon. These experimental long lines were the first applications of the atmospheric system that required multiple pumping stations and associated plumbing.

Before either long line was up and running, the House of Commons established a Select Committee to assess the "Merits of the Atmospheric System of Railway."[30] During March and April 1845, the dozen or so committee members peppered a dozen or so witnesses with more than two thousand pertinent questions. Beyond issues of safety, many technical questions were asked, most having implications as to whether the atmospheric system could be profitable. Among the people testifying were three Dublin & Kingstown officials, Joseph Samuda (Jacob had died the previous year in a freak ship accident), and railway managers and engineers including Stephenson, Cubitt, and Brunel.

It would be easy to frame the committee's hearings as a heated controversy between proponents of the atmospheric system (Brunel and others) and opponents (Stephenson and others), but this would be misleading. In fact, the hearings were surprisingly quite civil: witnesses did not confront each other and most agreed on the strengths, weaknesses, and economic uncertainties of the atmospheric system. The major axis of disagreement was whether atmospheric railways more than a few miles long (thus requiring two or more pumping stations) could be built at all, and if so whether they would yield a sufficient return on investment. In short, the hearings were a hard-nosed capitalist exercise.

No one dissented from the claim that the atmospheric system's most novel component, the valve, was effective. Stephenson even conceded "that the longitudinal valve is a complete triumph of mechanism. . . . I do not think anything could be more perfect."[31] There was general agreement that the atmospheric system was well suited to short lines with light trains running frequently; in this mode the steam engines driving the pumps could be used efficiently.[32] (Longer, heavier trains were best suited for steam power because locomotives could be added to increase the pulling power.) By running light trains often, an atmospheric railway might even make money, especially if the line had fares comparable to those of the much slower and bone-jarring horse-drawn omnibuses.

Although some atmospheric proponents—including Brunel—asserted that the system could be adapted to longer lines, the design of a scaled-up system would present many new problems, most of which were correctly foreseen by Stephenson and other critics. Beyond the difficulty of starting and stopping at intermediate stations was the near impossibility of running express trains. It was suggested that slower trains could be shunted onto a siding to allow an express train to pass, but that solution presented huge plumbing problems and would involve coordination by telegraph. Other problems would arise when the line had to cross roads and other rail lines. Some saw little hope that any crossing could be made on a level grade. Brunel conceded that the South Devon line would "avoid level crossings" and instead employ "bridges over and under the railway," and he acknowledged that this was a "much more expensive" solution.[33]

Because most short railways wouldn't remain short for long—they would be merged into larger steam networks—it would be difficult to make the connections. Thus the incompatibility of the steam and atmospheric systems would be a steep barrier to railway mergers. In addition, trains could not back up, a potential safety hazard; and if there was a breakdown on the line, getting the train running again could take many hours.

One especially vexing problem undercut the possibility of building long lines. To ensure that an atmospheric system's first costs were less than a locomotive system, Brunel and others did their calculations on the basis of a single track. By this time, however, many steam railways had two tracks—one in each direction—to accommodate more traffic and reduce head-on collisions. Stephenson pointed out, however, that running trains frequently in both directions on a single atmospheric line would severely limit the line's length, owing to the long delays caused by trains having to wait on sidings for an oncoming train to pass.[34] This conundrum implied three possibilities.

- Atmospheric lines would employ two tracks, causing first costs and operating costs to soar.
- Use of a single track would require many sidings and flawless coordination of trains by telegraph, inflating costs and raising safety concerns.
- No single-track, long atmospheric lines would (or should) be built.

In grappling with the technical and economic issues, atmospheric advocates offered untested ideas and argued bravely that the Croydon and South Devon lines would give engineers sufficient experience for solving the problems. Responding to what might have been sheer wishful thinking, Stephenson remarked that "I have no doubt that you may devise a plan, mechanically speaking, to overcome any mechanical difficulty, but the question is, is it feasible, and is it commercially useful."[35]

Stephenson, who had had more dealings with railway companies than any other engineer, knew that capitalists were most concerned about whether scaled-up atmospheric railways would be profitable. Members of the committee also knew this and pressed the witnesses for data about first costs per mile, operating costs per mile (fixed and variable), and so on. But witnesses presented so much inconsistent data, especially for hypothetical atmospheric lines, that major economic uncertainties remained.

The Select Committee concluded that the Dublin & Kingstown Railway "establish[ed] the mechanical efficiency of the Atmospheric power to convey with regularity, speed, and security the traffic upon one section of pipe between two termini." Moreover, relying on the testimony of Brunel, Cubitt, and another engineer, the committee was satisfied "that there is no mechanical difficulty" in using the system for "a line of any length."[36] The latter assertion, however, was not supported by the entirety of testimony given at the hearing. Anyone who read the published transcript, especially Stephenson's discussions, would have recognized that serious obstacles worked against long lines. Indeed, the two long lines still under construction, Croydon and South Devon—each with multiple pumping stations—turned out to be fiascos that sealed the atmospheric system's unhappy fate.

Engineer William Cubitt advised the London & Croydon to employ a single atmospheric system on a branch line.[37] The branch line opened in January 1846 but encountered many problems, beginning with steam engine failures and extending to badly leaking pipes. The valve was improved slightly to reduce wear on the leather and produce a more secure closure, but leakage continued and the valve seal melted in unusually hot weather.[38] A six-week pause to replace the valve didn't remedy every difficulty, but the system was working somewhat better.[39] However, the entire Croydon line required switching between atmospheric and locomotive sections, which caused inconvenience and delays.

Ironically, the problem that in the end doomed the Croydon atmospheric system was inflexibility in the face of increasing usage.[40] One would have thought that more passengers and more freight would have helped the bottom line, but this was true only if the line could increase capacity economically. As more and heavier cars were added to trains, the pumping stations had to work longer and harder to provide enough vacuum for the traffic—increasing the wear and tear on the valve, as well as on the steam engines and pumps, and ballooning the consumption of coal. The result was operating costs far above those of a locomotive line. Unlike a locomotive-driven railway, however, the capacity of the atmospheric system could not be expanded economically once its limit of two-way trains on a single track had been reached.

The Croydon line was acquired by another railway company that utterly lacked confidence in atmospheric propulsion. In May 1847 the new company terminated the experiment and diverted all traffic to locomotive tracks. The failure of the Croydon line cast doubt on whether any atmospheric line with multiple pumping stations and growing usage could ever turn a profit.

Brunel's Devon line was the atmospheric system's last hope.[41] If he succeeded, perhaps capitalists would take a second look at the radical technology. Having been impressed by visits to Wormwood Scrubs and Kingstown and Dalkey, Brunel asserted that an atmospheric line would be cheaper to build and more economical to operate than a locomotive-driven line. Although the South Devon line had been authorized in 1844 for locomotives on two tracks, Brunel recommended the adoption of a one-track atmospheric system; his cost estimates claimed a savings of £207,000 in construction. Despite some vocal opposition on the board, the company agreed to Brunel's proposal.

The railway was projected to run 52 miles from Exeter to Plymouth, accompanied by a telegraph line. Brunel detailed the specifications for the steam engines and they were ordered from Britain's best builders, including Boulton & Watt and Rennie & Maudslay. He also reached agreement with Samuda to purchase, install, and maintain the longitudinal valve. Clearly, South Devon would have the finest equipment, raising the odds for success. Pipes were ordered and construction began on the first twenty miles, which included eight pumping stations along the line's wide-gauge track. The steam engines, boilers, and pumps were installed in large buildings whose stately exterior disguised their industrial interior (Figure 3.3).

Construction was proceeding very slowly and this worried the company directors, especially after they learned that the Croydon line had been abandoned. Brunel blamed the delays on failure-prone steam engines, which "have proved sources of continued and most vexatious delays both in the unexpected length of time occupied originally in their erection and in subsequent correction of defects in minor parts."[42] The undependable steam engines made it

Figure 3.3. The South Devon atmospheric railway. Source: a painting by Nicholas Condy (1848). Note Dawlish pumping station at right.

impossible, according to Brunel, to complete testing of the line. Trials finally began early in 1847 with mixed results. By early fall, however, trains were running at 40–50 miles per hour. It was time, Brunel decided, to open the first fifteen miles of the line for daily service, even though the telegraph was not yet operating. On September 13 two trains began running in each direction between Exeter and Teignmouth.

The number and weight of trains was gradually increased and in mid-December the line was extended another five miles to Newton. Service for the most part was reliable, but the valve was showing signs of trouble. A heavy frost caused damage and delays. Although a large crew maintained the valve, the work did not prevent deterioration of the leather and even long cracks appeared. Another problem was rusting of the valve's iron parts, which had been stored for more than a year before installation. The leather too had sat unused for long periods, hastening decay and contributing to sections of the valve being torn as the piston carriage passed. And water sometimes leaked into the pipe. Despite continuing valve problems, the crew did keep the line running. To the consternation of Brunel and the directors, however, maintenance expenses were spiking. An entirely new valve might have been the solution.

An additional source of concern was coal consumption in the pumping stations: it was almost three times greater than Brunel's original estimates. Part of the reason was that the steam engines and vacuum pumps were working twenty hours a day to accommodate the traffic, and working very hard to overcome the persistent leakage. One modern researcher, Charles Hadfield, suggests that the steam engines were underpowered and the pumps and pipes too small for the demands placed on them.[43] Brunel could have recommended upgrading the entire system, but instead yielded to the critics and agreed to end the expensive experiment. Construction costs alone had more than doubled from Brunel's original proposal, reaching £433,991.[44] And total operating expenses were indefensibly high.

The South Devon atmospheric line was taken out of service in September 1848, its pieces sold for a loss of around £350,000, and the line converted to locomotive operation.

* * * * *

The demise of the South Devon dashed any hopes that Clegg and Samuda might have had for additional adoptions of the atmospheric system. Why did no other company adopt this passenger-friendly technology? Historian Henry Atmore suggests that "had the atmospheric worked perfectly perhaps it could have overcome the constraints on railway capital imposed by the collapse of the mania."[45] This is a curious claim because no *long* line seemed capable of working "perfectly" without incurring intolerably high first costs and excessive operating expenses compared with steam railways. The supposed "constraints on railway capital" were illusory: despite the burst of the speculative bubble and stock market crash of 1847, steam railways were being built at an accelerating pace. There was ample capital for railways, just not atmospheric ones.

Growth in the construction of steam railways clearly indicates that companies had made a deliberate decision to reject the atmospheric system. Given that its performance shortcomings and economic uncertainties had been widely publicized in railway journals, books, and the popular press during the second half of the 1840s, companies chose steam despite its many deficiencies in safety and passenger comfort. After all, the technological requirements and costs of steam railways were becoming fairly predictable and they promised a reasonable return on investment. In addition, their performance was improving dramatically (spurring profits), for they could carry heavier loads at higher speeds with greater reliability.[46] And they were easily expanded. In their choice of a propulsion technology, then, railway companies prioritized the bottom line. Rejection of the atmospheric system was in all respects a sound business decision.

The atmospheric railway was a boutique, high-maintenance technology that only worked reasonably well on very short lines with light trains running often—as on Kingstown and Dalkey (ended in 1854) and St. Germain (ended in 1860), both of which also depended on gravity in one direction. The constraints that worked against longer atmospheric lines had been known since the reports of the Board of Trade (1842) and the Select Committee (1845). Yet Croydon and South Devon, two long lines requiring multiple pumping stations and complex plumbing with no precedents, were still built.

The question is why did the companies take on these projects? The answer, I suspect, is the influence of prominent engineers. Working with a dearth of reliable data on the lines' likely costs and probable performance, William Cubitt and Isambard Kingdom Brunel crafted optimistic estimates of first costs and operating expenses that persuaded the Croydon and South Devon companies to adopt the technology. No doubt these respected engineers were enamored with the novelty and cleverness of the atmospheric system and the enormous creative challenges of making it work. It is unlikely that either company would have adopted the atmospheric system had it not been for the engineers' sterling reputations and apparently convincing proposals, which papered over the reality that these lines were risky experiments that could fail miserably.

Although the atmospheric railway was an abject failure, some of its components have survived. Pieces of the South Devon pipe can be seen at several British museums, including the Didcot Railway Centre Museum. And South Devon pumping stations, their handsome buildings repurposed, still stand at Torquay, Starcross, and Totnes.

NOTES

1 https://en.wikipedia.org/wiki/Surrey_Iron_Railway, accessed 25 February 2016.
2 https://en.wikipedia.org/wiki/Middleton_Railway, accessed 25 February 2016.
3 Smiles (1868) is an early and informative biography.
4 On the Hetton Railway, see Smiles (1868:208–209).
5 On the Stockton & Darlington Railway, see https://en.wikipedia. org/wiki/Stockton_and_Darlington_Railway, accessed 26 February 2016.
6 On the Liverpool & Manchester Railway, see Smiles (1868).
7 Smiles (1868:323). On the trials see also Ritchie (1846).
8 For numbers of railway casualties, see Ritchie (1846:362).
9 http://www.historyhome.co.uk/peel/railways/expans.htm, accessed 4 March 2016.
10 A brief overview of the atmospheric railway's history is Atmore (2004). For book-length treatments, see Clayton (1966) and Hadfield (1967). Unless otherwise noted, information on the atmospheric railways comes from these sources.
11 Atmore (2004:262).
12 Samuda (1841) provided information on the Clegg–Samuda system and the test at Wormwood Scrubs.
13 Atmore (2004).
14 I calculated these numbers from data in Samuda (1841:9).
15 Clegg's (1839) earlier pamphlet was written before the Wormwood Scrubs test and is of interest owing to his claims and overly optimistic cost estimates.
16 Samuda (1841:38).

[17] Quotes in this paragraph from Samuda (1841:24, 47).
[18] Atmore (2004:264).
[19] Quoted in Hadfield (1967:32).
[20] Atmore (2004:263–264).
[21] Smith and Barlow (1842:6).
[22] Hadfield (1967:43–44).
[23] Quoted in Hadfield (1967:44).
[24] Hadfield (1967:48).
[25] On valve maintenance see B. D. Gibbons (pp. 19–20) and T. F. Bergin (p. 73–74), in Select Committee (1845).
[26] "The Atmospheric Railway, II," *Saturday Magazine*, 2 March 1844, pp. 85–86. Bramwell (1899:322–325) offers unique details on the Kingstown and Dalkey line.
[27] T. F. Bergin in Select Committee (1845:72).
[28] "The Kingstown & Dalkey Atmospheric Railway," *Illustrated London News*, 6 January 1844, p. 16.
[29] Hadfield (1967:213–222).
[30] Select Committee (1845:iii).
[31] Stephenson in Select Committee (1845:106).
[32] Samuda in Select Committee (1845:8).
[33] Brunel quotes in Select Committee (1845:35, 36, 39).
[34] Stephenson in Select Committee (1845:123–124).
[35] Stephenson in Select Committee (1845:114).
[36] Select Committee (1845:iii).
[37] Hadfield (1967:116–142) gives details of the Croydon experiment.
[38] On the improved valve, see Samuda in Select Committee (1845:3, 174); Ritchie (1846:312–313).
[39] Atmore (2004).
[40] Hadfield (1967:141).
[41] On the South Devon line, see Brunel (2006:105–132), Buchanan (2001:103–112), and Hadfield (1967:143–176). See also Gregory (1982).
[42] Brunel quoted in Hadfield (1967:154).
[43] Hadfield (1967:173).
[44] Hadfield (1967:172).
[45] Atmore (2004:279).
[46] Ritchie (1846).

4

PUTTING ELECTROMAGNETISM TO WORK

The electric motor may be the most indispensable technology of the modern world. At home electric motors are commonly found in our washing machine, heater and air conditioner, dishwasher, vacuum cleaner, hair dryer, and many more appliances. In gasoline-powered cars, electric motors raise and lower windows, circulate air from the heater and air conditioner, operate windshield wipers, and of course start the engine. Electric motors also drive "diesel" trains, streetcars and subways, and submarines. Today our lives depend entirely on electric motors, but it wasn't always so. The earliest attempts to commercialize electric motors, which rely on the force field created by magnetism, all ended in failure.[1]

In addition to attracting pieces of iron, magnets—because of their seemingly magical properties—also attracted people. Some believed they could employ magnets in machines that moved continuously on their own, but no one could make a functioning machine. After all, this alluring idea rests on the scientific impossibility of perpetual motion.

* * * * *

The discovery of electromagnetism in 1820 laid the scientific foundation for employing the dramatic effects of magnetism. Hans Christian Oersted, a Danish professor of natural philosophy, adopted the hypothesis that electricity and magnetism are related. Combining research and teaching like many creative professors, he discovered electromagnetism while doing a demonstration in class. When Oersted connected a wire between a battery's positive and negative poles, a magnetic field instantly surrounded the wire, as indicated by the movement of a nearby compass needle. Oersted's momentous discovery—that the flow of electric current in a wire creates magnetism—thrilled natural philosophers on both sides of the Atlantic. Not surprisingly, experimenters set to work inventing devices that increased the power of electromagnetism. And there was rapid progress.

Frenchman André Marie Ampère learned that he could strengthen the magnetism by winding the wire into a spiral or coil. Building on Ampère's findings, William Sturgeon—a teacher in England—wound his wire on an insulated piece of iron (the core) shaped like a horseshoe, which greatly intensified the magnetism. Finally, Joseph Henry (who taught at the Albany Academy in New York) insulated the wire instead of the iron horseshoe. This made it possible to wind many layers of wire on the core, which multiplied the electromagnet's power. Henry's handiwork was the first essentially modern electromagnet. His largest electromagnets could lift more than a ton.

In 1831 Henry showed that electric motors were possible because of the electromagnet's unique capability: it can be turned on and off. He illustrated this potential with a miniature electric teeter-totter (Figure 4.1), a kind of rocking-beam device. It had two permanent bar magnets placed upright on a wooden platform with their north poles at the top. Between the bar magnets was a stand on which the armature—an electromagnet with two wires at each end—could pivot. Either pair of wires supplied current to the electromagnet when it dipped into a pair of tiny mercury-filled cups attached to the battery. When the electromagnet was energized, its north pole was repelled by one bar magnet while its south pole was drawn toward the other. As the armature rocked, it withdrew one pair of wires from the cups and inserted the second pair on the other side. This immediately reversed the poles of the electromagnet, which then rocked in the other direction. The rocking motion—about 75 times a minute—continued as long as the current flowed.

This elegant device showed that magnetism produces continuous motion when that motion itself reverses the electromagnet's polarity. This is the basic design principle of all battery-powered and other direct-current motors. Mechanical "pole changers" became known as commutators and are still found in some electric motors.

For Henry the battery-powered teeter-totter was a "philosophical toy," merely a demonstration device for the classroom; even so, he published a detailed account in the *American Journal of Science and Arts*.[2] Excited by Henry's report, experimenters in Europe and America incorporated its design principle into rotary and reciprocating motors able to drive machinery. With generous support from the Russian government, Moritz Jacobi assembled a large motor, installed it in a 28-foot boat, and with a dozen passengers cruised up and down the Neva River at about three miles per hour. And American Thomas Davenport, a Vermont blacksmith, powered tools with his motors.

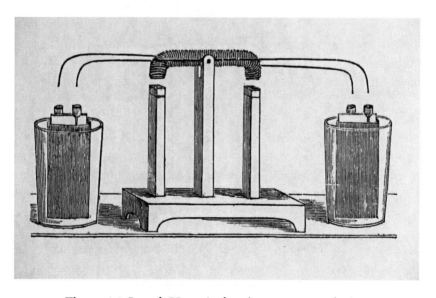

Figure 4.1. Joseph Henry's electric teeter-totter device.
Source: Henry (1831:342).

Although many experimenters showed that electric motors could drive machines and vehicles, Davenport mounted a uniquely ambitious effort to commercialize his motors. Around a decade later, Charles Page—a U.S. government patent examiner—built an electric locomotive.

Davenport learned that an electromagnet made by Henry was being used at the Penfield & Taft ironworks at Crown Point, New York. He traveled to Crown Point and, with $75 borrowed from his brother Oliver, bought the electromagnet. At home with his pricey purchase, Davenport carefully dissected it, forged a new iron core, wound the wire on it, connected a battery, and marveled that his electromagnet was now stronger than Henry's original.

In an autobiography Davenport recalled his epiphany. "Like a flash of lightening [sic] the thought occurred to me that here was an available power within the reach of man. If three pounds of iron and copper wire could suspend in the air 150 pounds of iron, what would three thousand pounds suspend?"[3] Davenport believed that "magnetic power" could replace other prime movers, especially steam engines.

Daily growing more important in an industrializing America, steam engines in the 1830s were beginning to power railroads as well as machinery in a few factories. And steamboats were already plying many rivers. Between Albany and New York City on the Hudson River, nearly two dozen steamboats ferried freight and passengers many times faster than travel by horse.[4]

Although steamboats were fast, their flimsy boilers invited disaster. In one 1833 accident on the Hudson, perhaps known to Davenport, the boiler of the steamboat *Ohio* exploded and killed five people.[5] Magnetic power, Davenport claimed, would be "a valuable substitute for the murderous power of steam . . . no more aching hearts & desolate homes occasioned by the awful spectacle of hundreds & thousands of human beings annually hurled into eternity." The power of electromagnetism, Davenport prophesied, "would no doubt eventually supersede that of steam." But first the visionary blacksmith would have to build powerful electric motors to hasten the transition.[6]

With the help of a neighbor Orange Smalley, Davenport built his first motor. The armature (the moving part) had four coils wound on the arms of an iron core shaped like the letter X. Surrounding the armature were arc-shaped permanent magnets attached to a circular wooden framework. A pole changer below switched the battery power among the coils, creating the alternating attractions and repulsions that rotated the armature.

Leading men of American science endorsed Davenport's motor and it was even described in European publications. He demonstrated it to Henry, who acknowledged Davenport's cleverness but insisted that electric motors could never replace steam engines because of the great expense of maintaining the batteries. In the batteries of the time, the zinc electrode gradually dissolved in the acid electrolyte and both had to be replaced—a considerable expense if the motor was used often or for long periods.

Davenport, now working with a cabinetmaker Ransom Cook, built more powerful motors (Figure 4.2), even a model railway, and hosted demonstrations for paying audiences. The meager funds these shows raised kept the project alive. Davenport and Cook also sought a patent, but their first try failed because a fire in the U.S. Patent Office in Washington destroyed the motor. A second attempt was successful, yielding in 1837 the first American patent for any electrical thing.

At the urging of a promoter and his lawyer, Davenport and Cook founded the Electro-Magnetic Association in 1837. They hoped to sell stock and use the money to build bigger motors. This was the earliest American company established to develop and market an electrical invention. With the first funds raised, Davenport and Cook set up shop in New York City, bought equipment, and hired helpers.

To promote stock sales, the company issued an impressive 94-page booklet offering shares at $100 each. They also obtained a glowing endorsement from Benjamin Silliman, a Yale professor who had witnessed several motors lifting weights. Newspapers around the country published stories about motors made by the blacksmith and cabinetmaker. Some accounts predicted that machines driven by electric motors would have revolutionary effects on civilization.

Despite the favorable publicity, stock in the Electro-Magnetic Association found few buyers. There was a good reason: the U.S. economy was suffering a severe depression following the financial

Figure 4.2. Patent model of Thomas Davenport's electric motor.
Source: National Museum of Natural History,
Smithsonian Institution, author's photograph.

panic of 1837. Also, the Electro-Magnetic Association earned an undeserved reputation as a stock swindle. Abandoned by Cook, Davenport was now working alone, continuing to construct more powerful motors—including one of nearly one horsepower that drove a Napier printing press. On this press he published two issues of the world's first electrical journal, the *Electro-Magnet and Mechanics' Intelligencer*. Despite these achievements, Davenport was garnering only limited funds from exhibiting motors. By 1842 he was nearly destitute—his stock worthless and no guardian angel in sight—and the disappointed inventor returned to Vermont and resumed his blacksmith trade, embittered that his electric motors had not conquered steam.

Davenport was not the only American enamored with the possibility of replacing steam engines with electric motors. Charles Grafton Page of Salem, Massachusetts, was a Harvard-trained physician who was disinclined to practice much medicine. His first love being electromagnetism, he invented dozens of little motors and other devices useful in classroom demonstrations. Page worked closely with Bostonian Daniel Davis, Jr., an important maker of scientific instruments. Offerings of Page's many electromagnetic inventions graced the pages of Davis's catalogs beginning in 1838.[7]

Perhaps Page's most important invention was what he called a "compound magnet and electrotome." It was in fact an induction coil, a kind of transformer that converted a battery's low voltage into high voltage through electromagnetic induction. Page's compound magnet consisted of two coils wound on a cylindrical iron core. First came the primary, which had relatively few turns, and then the secondary of many, many turns. (The ratio of turns between the primary and secondary coils determines whether the voltage is stepped up or down and by how much.) The electrotome caused the battery's direct current to pulsate automatically by opening and closing the primary circuit. The electrotome was necessary because unvarying direct current cannot induce a voltage in the secondary coil.

Given the many surprising effects of high voltage, the induction coil became an audience favorite in public lectures on electricity. Henry Morton of the Stevens Institute of Technology in Hoboken, New Jersey, had an induction coil made by Edward Ritchie, a Boston instrument maker. Weighing more than 335 pounds, its three-cell battery generated a 21-inch spark that could shatter a block of glass three inches thick. Viewers gasped in disbelief at this startling effect. Beyond impressing audiences, induction coils played an important role in the discovery of X-rays and radio waves in the late 19th century.

Page's "axial engine" was another invention that exploited electromagnetism. In an axial engine, an iron rod is alternately drawn into and then expelled from a coil; the reciprocating motion is controlled by a pole changer. Small axial engines became staples in Davis's instrument catalogs, but Page had a more ambitious motor in mind—one that would harness the rod's motion to a large machine.

In the meantime Page had moved from Massachusetts to Washington, where he was hired as an examiner in the U.S. Patent Office. There he developed a social network of important Washingtonians, including congressmen. And he continued to build larger axial engines, certain that his ingenious designs could make more efficient use of battery power.

In 1845 Page announced a new invention, an electromagnetic gun. The barrel had four coils arranged in a row such that an iron bar could pass freely through their hollow centers. The bar, carrying a wire connected to the battery, energized and de-energized the coils in succession as it passed through them. The gun could propel the iron rod as far as fifty feet.[8] With simple modifications the multi-coil gun could become a very powerful axial engine. Just such a motor would be employed in his next—and final—project.

Page wanted to demonstrate that a large axial engine could replace railroad steam engines. However, building a full-scale locomotive and a motor to power it was well beyond his means and expertise. He would need substantial funds for materials and skilled labor to design, cast, machine, and assemble the many parts. Counting on his connections, Page petitioned the U.S. Senate to support "an investigation of . . . applying electro-magnetic power to purposes of navigation and locomotion." This could, he suggested, result in "a general substitute for the dangerous agency of steam."[9]

Surprisingly, Congress a decade earlier had already set the precedent for supporting experimental electrical inventions when it voted $30,000 to fund Samuel Morse's telegraph line from Washington to Baltimore. A stunning success, Morse's project launched a frenzy of telegraph projects in the United States and around the world. Clearly, Page had every reason to believe that Congress might fund his pet project. After all, he was an international authority on electromagnetic devices, his expertise was undisputed, and he had an influential social network. Page even demonstrated one of his motors before Senator Thomas Hart Benson's Committee on Military Affairs, bloviating about the potential of electric power to do real work in place of steam.

On March 3, 1849, with no strong opposition, Congress approved legislation to grant Page $20,000. Apparently, Joseph Henry—now secretary of the new Smithsonian Institution—was not consulted about the feasibility of Page's locomotive project.

Page set up shop in the Washington Navy Yard in southeast Washington. Not only did he have access to people in the navy's departments of engineering, blacksmithing, and plumbing, but he also hired a machinist, laborers, assistants, and a chief engineer. However, Page spent most of the grant money on tools and supplies. Compulsive about ensuring that the materials going into the coils and mechanical parts were of high quality, Page used strength-testing equipment to assess samples from different vendors. That Page had organized his project around first-rate men and materials augured well for a good outcome.

Page and his crew set to work building ever-larger axial engines, including one that reached four horsepower. In assessing the operating costs of the big motors, Page acknowledged that they exceeded that of inexpensive steam engines, but he also claimed that "the expense was found to be less than the most expensive steam-engines."[10] Further, he maintained that only scaled-up experiments—not abstract calculations—could determine whether engines on the order of one hundred horsepower might turn out to be very efficient.

Despite Page's energy and optimism, work on the locomotive was moving slowly and funds were dwindling. Not surprisingly, he again petitioned Congress—seeking an additional $40,000 to continue the project. This time Page's petition ran into formidable opposition. Congressmen wondered, would every starry-eyed inventor with a sketch in hand petition Congress for a grant? And surely, added Senator Jefferson Davis, such appropriations were not sanctioned by the Constitution. The opposition carried the day and Page received no more federal money.

Embarrassed by his failure to complete the locomotive in a timely fashion, the beleaguered Page began pouring his own money into the project and then borrowed funds from friends. Months went by but still there was no vehicle. Finally, on April 29, 1851,

Page was ready to reveal his handiwork by putting the electric locomotive on the rails from Washington to Baltimore—the same route as Morse's first telegraph line.

The locomotive resembled a strange omnibus or coach (Figure 4.3), weighed 21,000 pounds, and carried seven passengers on its inaugural run. Two large axial engines, estimated by Page to be at least twelve horsepower each, provided power independently to two driving wheels five feet in diameter. The engine, a distant descendant of his magnetic gun, consisted of a series of coils that surrounded a heavy iron rod (five inches in diameter and four feet long) that moved back and forth. A sliding switch energized the coils, one after another, at just the right instant to add impetus to the moving mass of iron; the sequence was reversed on the return stroke. To ensure that the coils were activated on time, Page coupled the sliding switch to the iron rod.[11]

To motivate the motors Page constructed his own battery of one hundred cells; each cell was contained in two nested ceramic jars. On the way to Baltimore, the vigorous vibrations of the reciprocating motors began breaking the ceramic jars, reducing the power. By the time the locomotive limped toward Bladensburg—barely past the District of Columbia boundary—more than a dozen cells had

Figure 4.3. Charles Page's electric locomotive.
Source: *Greenough's American Polytechnic Journal* 4(1854):257.

failed and the situation was becoming dire. Declining battery power wasn't the only problem; the insulation on the motors also began to break down and Page had to disconnect several coils. With hope gone of reaching Baltimore, Page put the motors in reverse and the crippled locomotive returned to Washington. In all, the trip was interrupted seven times as the inventor attended to failing cells and arcing coils.

In a progress report to the secretary of the navy, Page put the best face on the "magnetic locomotive's" test run, which most others regarded as an abject failure. Obviously, no one could gainsay Page's claim to having made an axial engine as powerful as a steam engine; surely even larger ones were possible. Beyond the motor's technical competence, potential manufacturers and consumers would be interested in economy of operation. On this issue Page merely asserted, without supporting data, that "the larger my engines the greater the economy of power."[12]

Joseph Henry and other scientific authorities had previously argued that electric motors operated uneconomically compared with steam engines. Soon such arguments were buttressed with quantitative data. The calculations of Robert Hunt, a British professor of physics, were most influential. In his 1851 textbook, published the same year as Page's locomotive debacle, Hunt laid out an apparently compelling case against battery power. "A grain of coal burnt in the boiler of a Cornish [steam] engine, lifted 143 lbs. one foot high. A grain of zinc consumed in a battery to move an electromagnetic engine, lifted but 80 lbs. one foot high." Zinc, Hunt noted, was 24 times more expensive than coal per unit weight. Thus, he concluded, for an equal amount of work the operating expense of an electric motor would "be more than fifty times greater."[13]

Hunt lectured on battery power's economic liabilities at a meeting of the Institution of Civil Engineers in 1857. In discussions lasting the entire evening, no one dissented from his gloomy assessment. The consensus of the engineers was definitive. "[T]here could be no doubt . . . that the application of voltaic [battery] electricity . . . was entirely out of the question, commercially speaking."[14]

Three years later Hunt noted in another paper that zinc was now selling for £35 per ton, yet coal was less than £1 per ton. In an arresting—and perhaps not entirely facetious—conclusion, he observed that "it would be far more economical to burn zinc under a boiler, and to use it for generating steam power, than to consume zinc in a voltaic battery for generating electro-magnetic power."[15] For his studies of electric power, Hunt received the prestigious Telford Medal from the Institution of Civil Engineers.

Hunt's findings were given great credence—quoted often in technical and popular writings, extended by other scientists and engineers, and absorbed by potential motor investors and manufacturers. Indeed, such assessments mattered to businessmen and affected their investment decisions. The American industrialist and philanthropist Peter Cooper, who did much to encourage invention and technical education, was often beseeched by inventors to support their work on electric motors. For advice on this matter he relied on P. H. Vander Weyde, whose experiments led him to conclude that battery-powered motors were impractical. Accordingly, he advised Cooper to turn down the requests—which he did.[16]

Mature and rugged electric motors did not make it to market because manufacturers and investors, relying on the negative assessments of scientific authorities, declined to commercialize them. New York City, where Davenport and Cook had toiled, was a manufacturing hub—yet no company contracted to make their motors. Countless inventions competed for financial backing and many seemed much more likely than electric motors to turn a profit.

✳ ✳ ✳ ✳ ✳

With the benefit of hindsight, I have concluded that scientific authorities discouraged the commercialization of motors by inappropriately conflating science with issues of economy and consumer behavior. Blurring and extending the boundaries of scientific authority in this way gave men of science a sweeping mandate to pontificate.

In comparing electric and steam power, Henry, Hunt, and others focused on the expense of fuel (zinc versus coal) in producing equal amounts of mechanical power. Omitted from these calculations were the costs of buying and installing the power systems, acquiring the necessary space, hiring labor for operation and maintenance, and depreciation. At no time were the risks to human life factored in. And the estimates were context-free, unconstrained by the local factors that a proprietor would take into account in a specific installation—such as the amount of power needed or the anticipated frequency and duration of use. Also ignored was the larger issue of potential competition with non-steam power sources, such as human labor and (in later years) gas engines. These omissions seem especially glaring in analyses that were purportedly scientific. Indeed, by asserting the impracticality of electric power on economic grounds, scientific authorities were not doing science per se; rather, they were engaging in a crude kind of cost accounting. Nonetheless, these analyses would have appeared to be scientific because they were quantitative and because they were promoted by scientific authorities.

In the middle of the 19th century, when cost accounting was still in its infancy and formal marketing research did not yet exist, men of science expanded their social power in an understandable way—colonizing intellectual territories as yet unoccupied by other specialists. And having little other recourse, manufacturers and investors placed faith in their judgments. The conclusions of these scientific men seemed plausible to manufacturers and investors because all parties shared the belief that a businessman considering motive powers would be swayed by the narrowest economic considerations. Apparently, this was the case even though no laws of physical science prevented people from acquiring expensive new technologies.

Inventors such as Davenport and Page were not blameless in the electric motor's failure, for they committed a strategic blunder. If their arguments in favor of electric power had been only about small motors driving machines intermittently to replace human exertion, as in drills used by dentists and jewelers, scientific authorities might have framed their analyses differently and judged battery power useful—perhaps even economical—for certain applications. Instead inventors challenged the steam engine, invoking the emotionally salient boiler explosions that routinely killed many people at once. Evidently, targeting the demon of steam was a fatally flawed strategy because it set the agenda for the economic analyses of scientific authorities.

In the late 1870s, steam-powered dynamos began to produce electricity that was vastly cheaper than that produced by batteries and scientific authorities could at last endorse electric power. As a result, a host of firms developed and brought to market an endless variety of sturdy motors capable of continuous duty. According to economic historian Malcolm MacLaren, "[B]y 1887 there were fifteen well-known manufacturers of small [electric] motors in the [United States] who had produced over 10,000 motors of fifteen horsepower and below."[17] Following the lead of Davenport and other inventors, these manufacturers commercialized rotary motors; Page's axial engines had been shown by simple calculations to be ghastly inefficient.

One of Joseph Henry's largest electromagnets, which he built for Yale University, is on display at the National Museum of American History, Smithsonian Institution. However, his original teeter-totter no longer exists.

No traces of Charles Page's electric locomotive survive. However, many of his small electromagnetic devices (commercialized by Daniel Davis, Jr.) can be found in science and technology museums, including the National Museum of American History. This institution also holds the patent model of Davenport and Cook's electric motor, a most precious possession, as well as a circular track several feet in diameter on which a wheeled motor ran (the model train?).

In closely examining the patent model (Figure 4.1), I noted several points of interest. First, the commutator (the twisted wires between the four wooden supports) was poorly designed and easily damaged. Second, the motor has more parts than the patent drawing indicates. Affixed to the vertical rotor is a small gear that engages a much larger gear. The latter in turn drives a horizontal shaft whose other end is supported by a wooden pillar attached to the base. There is also a slot in the base, parallel to and directly below the shaft. These features would have allowed Davenport to demonstrate the motor's lifting ability: a string would be threaded through the slot, its upper end fixed to the horizontal shaft and its lower end attached to a weight. Connected to a battery, the motor would have wound the string and lifted the load.[18] No doubt an impressive demonstration.

NOTES

[1] Unless otherwise noted, information in this chapter comes from Schiffer (2008a).

[2] Henry (1831).

[3] Davenport (1851:5).

[4] Information in this paragraph comes from Redfield (1833).

[5] Redfield (1833:317).

[6] Quotes in this paragraph are from Davenport (1851:6, 6–7).

[7] On Page, I consulted Post (1976).

[8] Page (1845).

[9] Page, quoted in Post (1976:82).

[10] Page (1973a:412).

[11] On the motor's design and performance, see Post (1976:96); "Electro-Magnetism as a Motive Power," *Scientific American* 7(1851):64, 68; "The First Locomotive That Ever Made a Successful Trip with Galvanic Power," *Greenough's American Polytechnic Journal* 4(1854):257–264.

[12] Page (1973b:425).

[13] Quotes in this paragraph are from Hunt (1851:308).

[14] For synopses of Hunt's paper and ensuing discussions, see *Engineer* 3(1857):325, 364–365. Quote on p. 364.

[15] Hunt (1860:100).

[16] Vander Weyde (1886).

[17] MacLaren (1943:92).

[18] Schiffer (2013a: 89–90).

5

AUDACIOUS ENGINEER

Isambard Kingdom Brunel and the *Great Eastern* Steamship

Three new technologies emerged during the extraordinary decade of the 1830s: oceangoing steamships, steam-powered railroads, and electric telegraphs. These technologies enjoyed rapid adoptions, especially in the industrializing West, and accelerated the expansion of commerce and travel within and among countries. In the following decades, continuous improvements in many performance characteristics of these technologies made possible reliable and far-flung transportation and communication networks.

On occasion a new steamship, railroad, or telegraph was forecast to bring about revolutionary changes yet failed to meet expectations. Such was the fate of Isambard Kingdom Brunel's *Great Eastern*, an immense steamship that was projected to dramatically hasten travel between Britain and its largest eastern colonies, India and Australia.

Isambard acquired an interest in engineering at an early age. His father Marc Brunel was a British inventor and civil engineer of appreciable accomplishments. The elder Brunel steered his son into civil engineering and set high expectations. Sent to France, Isambard received mathematical and technical training and apprenticed with master craftsman Abraham-Louis Breguet. Isambard's confidence and ambition grew as he gained skills, for he aimed to create nothing less than great works. Often these great works were carried out at substantial expense, far more than the original estimate.[1]

✳ ✳ ✳ ✳ ✳

Residing in the port city of Bristol, Isambard Brunel was on good terms with the financial class and began securing commissions. At age 24 he won the 1831 competition to design a bridge over the Avon River at Clifton Gorge. In what would become his trademark move, Brunel's design was bold: an iron suspension bridge having a single span of 630 feet—an unprecedented length.

At an early age, Brunel expressed frustration with train travel. Writing about his first ride, he complained about the shaking of the Manchester Railway. "The time is not far off when we shall be able to take our coffee and write whilst going noiselessly and smoothly at 45 miles per hour. Let me try."[2] Soon he had the opportunity to try because Bristol's men of consequence wanted a railway link to London; they formed a committee and raised money to survey a route. Brunel received the commission despite insisting, "The route I will survey will not be the cheapest—but it will be the best."[3]

Brunel's route—which he surveyed on horseback—was as nearly level as possible, allowing rapid travel. He also envisioned it as the foundation of a more extensive railway network in southwest England. As Brunel had warned, the Great Western Railway (GWR), his grandiose name for the line, would not be cheap. His estimate for the 116-mile line—including bridges, viaducts, tunnels, stations, tracks, and locomotives—was a breathtaking £2.5 million.

In collaboration with a London committee, the Great Western Railway Company was formed and shares were offered to the public. The next step was to convince Parliament that the design was feasible, that sufficient shares had been subscribed to finance the project, and that protesting landowners had been mollified. After initially rejecting the project, in 1835 Parliament approved an amended bill for the Great Western Railway.[4]

Brunel was appointed the project's chief engineer, earning a princely annual salary of £2,000; subordinate engineers earned from £150 to £350, depending on experience; ordinary workers no doubt made less than £50.[5] To carry out the dozens of construction projects required by the GWR, Brunel hired—and brutally micromanaged—many subcontractors, visiting construction sites in his well-appointed britzska, a hearse-like coach.

In the early 19th century, the engineering sciences for canals, water systems, harbors, bridges, locomotives, and so forth were advancing rapidly. As a result, engineers increasingly specialized in one or just a few technologies. But Brunel sometimes tackled technologies whose engineering science he had not mastered. He could have adopted the speedy and durable locomotives that Robert Stephenson and his son, the foremost railway engineers at that time, were building. Believing he could do much better, however, Brunel designed the first generation of GWR locomotives. Unfortunately, they did not serve well; fortunately, Brunel hired the very young Daniel Gooch, who proved himself a capable builder of Stephenson-style locomotives, as locomotive superintendent.

Constructing the many tunnels, bridges, and viaducts posed challenge after challenge—especially the 1.75-mile-long Box Tunnel, then the longest in Britain. In these domains of civil engineering Brunel was at his best, his innovative designs usually succeeding.

While the public and company directors anxiously awaited completion of the entire GWR line, segments were opened as soon as they could carry traffic. At last, on June 30, 1841, the first train chugged all the way from Temple Meads Station in Bristol to Paddington Station in London (Brunel had designed both stations). The line's total cost was £6,282,000—2.5 times Brunel's estimate.[6]

During the GWR's construction, Brunel built shorter railways that connected to the Bristol–London line, but a far more radical connection was in the offing. At a meeting of the GWR board, a director objected to Brunel's plans, claiming the railway was too long. Brunel quipped that it wasn't long enough, for the line should reach all the way to New York City.

After the meeting Brunel and close friend Thomas Guppy, a wealthy Bristol engineer, discussed building a trans-Atlantic steamship.[7] When this conversation took place in 1835, no ship had crossed the Atlantic entirely under steam power. Optimistic that steamships could attract many passengers and perhaps cut travel time in half (sailing usually took 4–8 weeks), Guppy, Brunel, and a group of Bristol investors formed the Great Western Steamship Company (GWSC).

To move the project along, Brunel set up a committee consisting of himself and Guppy; Christopher Claxton, an experienced naval captain; and William Patterson, owner of a Bristol shipyard. Acknowledging that they needed to tap additional expertise in marine architecture, committee members visited shipyards and—at the Woolwich Royal Navy Dockyard—obtained advice and drawings from Admiralty engineers.

An important obstacle that had to be overcome was the belief that a steamship carrying enough coal for the voyage would have room for neither passengers nor cargo. Brunel, who now commanded the engineering science of marine architecture, used simple calculations to show that this belief was wrong. What's more, he knew that the larger a ship (and its corresponding coal stores), the more cargo and passengers it could accommodate. He also determined that a longer ship would be faster, more fuel efficient, and smoother through the water.

Accordingly, the *Great Western* steamship was 212 feet long—a record that stood for just one year. It was propelled by a paddle wheel on each side of the iron-strapped oak hull. To slow the buildup of barnacles and prevent shipworms (*Teredo Navalis*) from munching the hull, vulnerable areas were covered with copper sheets. The two steam engines were massive: built by Maudslay Sons & Field to Brunel's specifications, their cylinders were 70 inches in diameter and had an 84-inch stroke.[8] Each engine was directly coupled to a paddle wheel.

The *Great Western* was built in William Patterson's Bristol shipyard along the "floating harbor." Fed by the Avon and Frome rivers, the floating harbor was a long basin in the old channel of the Avon River that had been dammed.[9] Of course the only thing that actually floated in the harbor was a ship, which could pass through the Cumberland Lock on its way to the ocean, about eight miles away.

Construction went smoothly and the *Great Western* was launched on July 22, 1837. A month later, after acquiring sails for its four masts, the vessel headed to London for fitting out, where the engines and interior furnishings were installed (Figure 5.1). In its first trials under steam, the *Great Western* cruised swiftly at eleven knots (one knot = 1.15 miles per hour).[10]

To the north of Bristol, Liverpool merchants also wanted to create a trans-Atlantic steamship service. They formed the British and American Steam Navigation Company, commissioned a ship, and expected it to depart before the *Great Western*. However, construction delays thwarted the Liverpool company's plans. Meanwhile, the *Great Western* was poised to make the crossing first. Attempting to upstage the *Great Western* on its inaugural voyage to New York, the Liverpudlians took drastic action and remodeled the *Sirius*, an Irish ferry that had run between Cork and London.[11]

Filled with coal, the *Sirius* left for New York four days before the *Great Western*. Out of coal and burning just about anything

Figure 5.1. The *Great Western*.
Source: Nicholls and Taylor (1882:311).

flammable, the *Sirius* arrived at New York Bay but—having failed to take on a pilot—ran aground. A few hours later, on April 23, 1838, the *Great Western* steamed to its New York dock on the East River with more than a hundred tons of coal still in its bunkers. The *Great Western's* historic voyage of 15.5 days, averaging 8.66 knots, introduced regular steamship service across the Atlantic.[12]

The *Great Western* made dozens of two-way crossings during eight years of service between England and New York. After sale to another company, the ship ran between England and the West Indies. The steamer's last voyages were as a troopship during the Crimean War.[13]

When the GWSC directors decided to build a sister ship for the trans-Atlantic route, they turned to Brunel and Guppy. Instead of patterning it after the *Great Western*, the team arrived at a more radical design. The *Great Britain* combined two new technologies—iron hull and screw propeller—never before employed together in one vessel. And at 322 feet the *Great Britain* would be immense, the largest ship ever built.[14] It would have six masts whose sails could be used for auxiliary propulsion and maintaining stability in storms. The ship was also slated for four decks, a crew of 120, and 360 passengers.[15] Brunel estimated that the *Great Britain* could be built for £70,000.

Construction of the *Great Britain* began in mid-1839 but proceeded in fits and starts because the hull and engines had to be modified to permit screw propulsion.[16] The huge power plant consisted of two twin-cylinder steam engines built at Bristol. Each cylinder was 88 inches in diameter and had a six-foot stroke. The crankshaft of the tandem engines revolved at eighteen revolutions per minute, too slow for the propeller, so it was geared up through two toothed drums—a large one on the crankshaft, a small one on the driveshaft—connected by a chain. The driveshaft and thus propeller revolved at 53 revolutions per minute.

The mid-stream modifications not only caused delays but also cost overruns that strained GWSC's finances. And more difficulties lay ahead. The *Great Britain's* beam was 50.5 feet, slightly larger than the width of the Cumberland Lock. Dock officials had assured Brunel that the lock would be widened in time to launch the big ship, but bureaucratic hassles were not resolved until the end of 1844. GWSC at last received permission to modify the lock and Brunel's crew tore down upper courses of the masonry to ease the ship's passage. Finally, during the last high tide of the season, the *Great Britain* squeaked through the Cumberland Lock.[17]

After sea trials that reached twelve knots, the *Great Britain* headed to London for fitting out (Figure 5.2). On the vessel's inaugural voyage, which began on July 26, 1845, the *Great Britain* crossed the Atlantic in an impressive 14 days, 21 hours. During the next few crossings, there were problems with the Brunel-designed propeller and in heavy weather on one trip a mast was lost. The ship was also slower than expected and had a tendency to roll. To solve these problems, the GWSC undertook a series of modifications and returned the ship to service. And then disaster struck.[18]

Working from old charts on a trip west, the *Great Britain*'s Captain Hosken mistook a new lighthouse for an older one. Steaming ahead, the ship ran aground in Dundrum Bay on the coast of Ireland. At Brunel's insistence, the following year the *Great Britain* was refloated and sailed again. However, a 67 percent cost overrun for construction, expensive modifications in service, and £34,000 to

Figure 5.2. The *Great Britain*.
Source: Claxton (1845:Frontispiece).

refloat the beached behemoth sent the GWSC into bankruptcy. The *Great Britain* was sold to Gibbs, Bright & Company for a paltry £25,000.[19]

The new company gave the *Great Britain* a complete makeover—replacing the engines, reinforcing the hull, changing the gear mechanism, and so on. After brief trans-Atlantic service, the ship was sold again and modified for the Australia run, made popular by the discovery of gold there in 1851. The *Great Britain* ran this route successfully for almost three decades.[20]

In 1851 William Hawes—chairman of the Australian Royal Mail Steamship Company—asked his old friend Isambard Brunel to advise the company, which was keen to commission a pair of steamships for travel between England and Australia. Brunel developed the general specifications and requested that John Scott Russell bid on the project. Russell, whom Brunel had known for many years, owned a well-equipped shipyard (including a foundry) at Millwall on the Thames in London.[21] Russell was well known in scientific circles because of his wave-line theory, which specified a hull's contour such that it minimized the vessel's drag in the water.[22] He also discovered "waves of translation" (now called solitons); formed by a ship's bow, they outrun the ship and maintain their shape over long distances.[23]

Russell secured the commission and worked up designs for two iron screw-driven steamers—the *Adelaide* and the *Victoria*—both almost as large as the *Great Britain*. Construction went very rapidly and the two ships were completed by the end of 1852. A four-month round-trip by sail could now be done in half the time. Brunel was pleased that his support of Russell had turned out so well.[24]

On their round-trips to Australia, the *Adelaide* and the *Victoria* had to stop at a coaling station in Cape Town, South Africa, to replenish their bunkers. No steamship at that time could make the arduous voyage around Africa, across the Indian Ocean, and on to Australia with enough coal for a nonstop voyage—much less a return trip. This is one reason why sailing ships, some with auxiliary steam propulsion, were often used on the route. Pondering the time wasted and great expense of coaling at Cape Town, Brunel decided to draft a sketch for a steamship large enough to make the Australia run and return without stopping for coal.[25]

Its working name would be the Great Ship, though others called it Leviathan; eventually the ship became, predictably, the *Great Eastern*.[26] Brunel worked closely with Russell on the ship's design: it would be twice as long as the *Great Britain*, have five times the displacement, and carry four thousand passengers in three classes. Russell calculated that the ship would require an engine of 8,500 horsepower to reach Brunel's target speed of fourteen knots. Because no single engine could deliver anywhere near that much power, Brunel suggested they use both screw and paddlewheel propulsion with separate engines.

The first task was to find a company willing to commission the ship. Russell advised Brunel to approach the Eastern Steam Navigation Company, where his powers of persuasion would come in handy since "some of the most astute financiers easily fell under his spell."[27] That Brunel had become a daring and successful marine architect and Russell a highly skilled shipbuilder gave their proposal exceptional credibility. A majority of the company's board accepted the Brunel–Russell proposal.

To finance the project, the company sold shares. At that time the sale of shares—called a subscription—required no money up front; it was simply the buyer's commitment to pay when the company made a "call" on the shareholder. The project's total capitalization (up to £20 per share) was £800,000, but it was allowed to proceed when paid-up calls totaled only £120,000. Both Brunel and Russell bought many shares.

Taking an even greater risk, Russell also agreed to accept payment in company shares. The company's largest backer was Charles Geach who supplied iron for the ship in exchange for shares. Contracts were signed on December 22, 1853, and work began. Russell was building the hull and paddle engine, James Watt & Company the screw engine.

Although Brunel and Russell worked together amicably on the *Adelaide* and the *Victoria*, their relationship on the *Great Eastern* project grew fraught. As the project engineer, Brunel had the right to approve all detailed drawings but his endless suggestions for minor changes became a burden.[28] In addition to micromanaging Russell and his skilled workers, Brunel badgered the builder about supposedly missing materials and construction delays often caused by Brunel's own tinkering with the plans.

Adding to the tensions, Geach—who had subscribed two thousand shares—died in 1854, stressing Russell's finances. Worse still, Russell's shipyard suffered two fires that his insurance did not fully cover.

According to Russell's biographer George S. Emmerson, work progressed despite the increasing financial pressures. Three million rivets—heated, held, and hammered by hand—joined thirty thousand wrought iron plates 10 feet long and 33 inches wide in an over-under-over pattern. Cut, drilled, and shaped by steam-powered machines, the plates had to be hoisted into place with ropes and pulleys because Brunel refused Russell's request for a traveling crane. Despite the many challenges, the ship's "giant form began to rise like a rust-colored primaeval monster with platers, riveters and their helpers swarming like ants over the bare bones and armoured hide."[29]

Owing to the ship's unprecedented size (692 feet long), it incorporated important safety features. Following the *Great Britain*, *Adelaide*, and *Victoria*, the *Great Eastern* had transverse and longitudinal bulkheads that divided the ship into watertight compartments (Figure 5.3a). As long as doors were closed, a potentially catastrophic breach would be confined to one or two compartments. In addition, the ship had inner and outer hulls separated by two-foot-wide iron plates riveted in place with angle irons and gusset plates (Figure 5.3b.). If the outer hull were breached, the ship would remain afloat—a feature that would have been useful on the *Titanic*, sunk by an iceberg's glancing blow.

The paddle engine had two pairs of cylinders—each with a 74-inch bore and 15-foot stroke—that connected to the 40-ton crankshaft (Figure 5.4a).[30] The screw engine was also huge, as required to rotate the 24-foot-diameter propeller (Figure 5.4b). Supplying steam to the engines were 10 boilers and 112 furnaces.[31]

The ongoing construction of the *Great Eastern* was a public spectacle that attracted thousands of visitors who paid to view the

Figure 5.3a. The *Great Eastern*'s transverse bulkheads under construction. Source: courtesy of National Museum of American History, Smithsonian Institution, photograph by Robert Howlett.

Figure 5.3b. The *Great Eastern*'s double hull under construction. Source: courtesy of National Museum of American History, Smithsonian Institution, photograph by Robert Howlett.

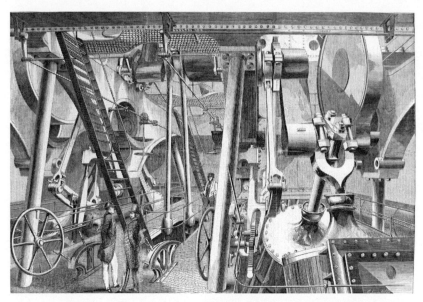

Figure 5.4a. The *Great Eastern*'s paddle wheel engine.
Source: author's collection from the *Illustrated London News*, ca. 1858.

Figure 5.4b. The *Great Eastern*'s screw engine.
Source: author's collection from the *Illustrated London News*, ca. 1858.

ship at close range; even Prince Albert, Queen Victoria's consort, showed up. Many magazines and newspapers in Britain and abroad chronicled the work. The *Illustrated London News* often visited Russell's Millwall yard, where it documented the ship's progress with text and stunning photograph-based drawings.

Late in 1855 Russell's financial difficulties passed the critical point. He was forced to liquidate his company and discharge his employees.[32] Work on the big ship stopped. However, the Eastern Steam Navigation Company—the legal owner of the unfinished vessel—had little choice but to rehire Russell to complete the job. Arrangements were made with the new owner of the shipyard and its equipment and construction resumed with Russell as the project's "Shipbuilder and Engine Maker." Russell was again under Brunel's supervision.

Money was still tight because many shareholders, including Russell, were in arrears on their calls for £192,000.[33] Even so, a launch date was set for October 1857. But this date slipped because, among other delays, Brunel insisted on many rehearsals of the unusual launching procedure.

Because the Thames at Millwall was barely wider than the length of the ship, the vessel could not be launched bow first, the usual practice. This is why the *Great Eastern* was built slightly above but parallel to the Thames. Brunel and Russell differed on how to ensure a successful sideways launch. To assuage his fear that the ship might move dangerously fast down the 1-in-12 slope, Brunel insisted on a highly controlled launch with enormous chains restraining the vessel. Russell believed that the ship would slide gradually into the water without restraints. Brunel also preferred iron for the sliding surface on both ship and launch ways, but Russell insisted that proper lubrication of a wood-on-wood interface would promote a smooth launch. Not surprisingly, Brunel did not defer to Russell's expertise but proceeded with his own launching plans.

Having paid for the privilege of standing on the opposite bank, a large crowd witnessed the launch on November 3, 1857. Also attending (but not paying) were distinguished visitors, "including the Duc D'Aumale, the Count de Paris, the Bavarian Minister, the Siamese Ambassadors, the Lords of the Admiralty," and London city officials.[34] Perched on a launching platform (Figure 5.5),

Figure 5.5. Brunel directing the launch of the *Great Eastern*. From left to right: John Scott Russell, Henry Wakefield (Brunel's assistant), Isambard Kingdom Brunel, and Lord Derby. Source: Courtesy of Wikimedia Commons; photo by Robert Howlett.

Brunel gave the order for the hydraulic rams to push the ship forward. However, progress stopped after just a few feet. The ship did not budge farther because, as Russell had feared, the iron was binding on itself. The launch did not resume until Brunel had obtained 21 of the most powerful hydraulic rams in England. Nudged inch by inch over many weeks by the rams that didn't explode, the *Great Eastern* finally reached the water on January 31, 1858.[35]

❊ ❊ ❊ ❊ ❊

Not only did the launching fiasco deeply embarrass Brunel, but it also cost about £100,000, put the company in dire financial straits, and compelled the project's reorganization. A new company, the Great Ship Company, was formed and bought the ship for £160,000. Stockholders in the old company received shares in the new one at a highly unfavorable ratio. The new stock offering raised enough money to complete the ship and Russell was awarded a £125,000 contract for the *Great Eastern*'s fitting out.[36]

Although much work remained, on August 8, 1859, "Russell held a banquet aboard the ship to celebrate its 'completion.' "[37] Before the ship's first sea trial in September, however, Brunel suffered a massive stroke and died ten days later on September 15. He would be spared the *Great Eastern*'s further indignities.

The sea trial in the English Channel began with high hopes. Lecturers on deck "were discussing the safety and prospects of the ship, when suddenly . . . up shot the great forward funnel of the ship in two pieces, thirty feet in the air, amid a shower of splinters and pipes and a volume of steam and smoke."[38] Although damage was confined, several firemen were scalded to death. The cause of the explosion was a valve in a Brunel-designed funnel heater that had been inexplicably closed. After repairs but before the ship's inaugural voyage, Captain Harrison drowned while paddling ashore. Added to the interminable delays and bankruptcies, these unfortunate accidents—and Brunel's death—fostered the belief that the *Great Eastern* was jinxed.

The Great Ship Company's directors planned to put the *Great Eastern* on the trans-Atlantic route, believing that it would soon be a paying proposition. By this time, however, 45 vessels—following the lead of Brunel's *Great Western*—already served the North Atlantic route.[39] Clearly, the *Great Eastern* would face formidable competition on a route for which it had not been designed. Smaller vessels, by the way, spent less time in port, thus making more trips per year and potentially increasing profit.

At a meeting of the London Society of Arts, several speakers argued against the North Atlantic route. In particular, William Hawes—seemingly channeling Brunel—argued strongly for heading east, noting that the ship was designed for travel in warm climates and no provision had been made for heating the cabins and saloons. Trade with India was robust and the *Great Eastern* would likely turn a tidy profit if it traveled this lucrative route. In a telling comment, Hawes lamented that "other interests than those of the shareholders, and, perhaps, the desire of exhibiting this great work to our transatlantic friends" were behind the decision to go west.[40] Hawes's arguments were ignored. After several postponements of the departure date, the *Great Eastern* would be going to New York, a mobile advertisement for Britain's industrial might.

With light-colored paint on five funnels and six masts, a white stripe on the black hull, and a crew of four hundred, the *Great Eastern* set sail on its inaugural voyage from Southampton to New York on June 17, 1860—more than six years after construction began (Figure 5.6). The vessel arrived at Sandy Hook, southern gateway to the New York harbor, eleven days and two hours later.

According to the *New York Times* correspondent on board, the voyage "demonstrated the *Great Eastern*'s superiority as a sea-

Figure 5.6. The *Great Eastern* begins its inaugural voyage.
Source: *Illustrated London News*, 23 June 1860, pp. 600–601.

SPECTACULAR FLOPS

going vessel, the excellence and reliability of her machinery, and her unrivaled safety and remarkable comfort as a passenger steamer." The ship's sail-aided top speed of 14.5 knots achieved Brunel's performance goal despite occasional "rough weather" and the engines being run conservatively at less than full steam pressure.[41] Although impressed by the *Great Eastern*'s mechanical prowess, the *Times* correspondent joined a chorus of critics questioning whether commercial success was possible. After all, on this voyage the ship carried only 38 paying passengers.

After some sprucing up at a New York dock, the ship was thrown open to the paying public for $1 a head. A saloon was provided on the upper deck to accommodate "the hungry and thirsty multitude."[42] Deemed excessive by the press, the charge was lowered and visitors flocked to the spectacle, up to eighteen thousand on one day.

Company directors found additional ways to use the ship's notoriety to generate revenue. A two-day excursion to Cape May, a resort community in New Jersey where the Delaware River meets the Atlantic Ocean, attracted about 1,500 paying customers. However, the trip was a public relations disaster. The press pounced, reporting that customers were sold poor food, poorly served, in an overcrowded saloon; at night they could rent a mattress (fifty cents) and bed down on the deck. Evidently there was a management problem.

With a stop at Hampton Roads, Virginia, the *Great Eastern* next ventured up the Chesapeake Bay to the Naval Academy at Annapolis, Maryland, where President James Buchanan came aboard.[43] Treated to a sumptuous feast, the president raised a toast to Queen Victoria. Upon the ship's return to England, the queen "bestowed knighthoods all around."[44] Manifestly, the *Great Eastern* had become a monumental national symbol, representing "the power and majesty of the greatest and most extraordinary nation that ever stamped its impress on the world."[45]

Before the second voyage to the United States, Abraham Lincoln was elected president and almost immediately southern states seceded from the Union. As a precaution, Britain chartered the *Great Eastern* to transport more than two thousand troops, their families, and two hundred horses to Canada.[46] More voyages to the

United States took place during the Civil War but they attracted too few paying customers and too little cargo to be profitable. However, incidents on two trips proved that the *Great Eastern*—though a money pit—was at least seaworthy under the direst conditions.

The shipping lanes leading to New York harbor concealed dangerous obstacles. When the *Great Eastern* struck the rocky bottom on Long Island Sound, the result was an eighty-foot-long gash in the outer hull. This damage caused a slight listing but the inner hull kept the vessel afloat. A temporary repair was done in New York at great expense, and a permanent one in Liverpool at even greater expense.

On another voyage to New York, the ship's durability was tested by a ferocious storm that struck on the second day. One paddle wheel was torn off, the other damaged beyond use. Worse still, the damaged rudderpost left the rudder swinging freely. With a jury-rigged rudder, the ship limped to Queenstown, Ireland, on propeller power alone, where the ship was repaired. Aboard during the aborted voyage, famed author Jules Verne vividly described the troubles in his novel *Une Ville Flottante* (*A Floating City*).

In 1863 the *Great Eastern* made three final trips to New York, carrying a total of just 3,695 passengers for a net loss of £20,000.[47] Its finances badly battered, the Great Ship Company entered bankruptcy in January 1864. The white elephant of the sea was put up for auction but fetched only £25,000. The buyer was Brunel's old associate Daniel Gooch, who along with Thomas Brassey and John Pender founded the Great Eastern Steamship Company.

Gooch believed the *Great Eastern* could be used to lay a new trans-Atlantic telegraph cable. The cable had been the pet project of American promoter Cyrus Field, who raised most of the money for it in England and secured in-kind contributions from both governments. Laid by two ships in 1858, the old cable was poorly designed and stopped working after a few weeks of dreadful service. Gooch reasoned that loading the heavy cable on one ship would simplify the laying process, and that success would lead the way for other cable projects around the world. The new company negotiated a favorable deal, leasing the ship to—and acquiring a financial interest in—the cable-laying conglomerate.[48]

The *Great Eastern*'s first attempt to lay the redesigned cable failed after the cable broke, the fault of the paying-out machinery. Field raised more money and, after the machinery was replaced in 1866, the second attempt succeeded. The cable worked well, transmitting ten words per minute in Morse code, just as Field had promised. Immediately, the Atlantic telegraph was highly profitable. During the next several years, the *Great Eastern* laid several long submarine cables, including one from Aden to Bombay (Mumbai).

During its second life as a cable ship, the *Great Eastern*—which had caused so much financial pain to so many companies and investors—helped create great wealth. But the aging vessel was replaced in 1874 by the *Faraday*, a special-purpose cable ship.

The rest of the *Great Eastern*'s life was a sad decline. After 1874 the ship never returned to deep water and at one point was docked at Milford Haven for a dozen years. It was employed a few times in money-losing and aborted schemes and auctioned again. As a floating amusement park on the Mersey River, the ship represented Lewis's department store with a vast music hall, stalls, and sideshows. After this indignity the *Great Eastern* was displayed briefly in Dublin. Finally, in 1888 Brunel's brainchild was sold for scrap to Bath & Company, realizing a mere £16,500. In the end the Liverpool shipbreaker got no bargain because dismantling the floating fortress took twice as long as estimated.[49] When the *Great Eastern*'s life ended in 1889, there was no larger ship in the world.

Some writers have suggested that the *Great Eastern* flopped as a passenger liner because it was "constructed years ahead of her true era."[50] I hasten to point out that "built before its time" is a classic non-explanation. It is true that gigantic passenger liners such as the *Lusitania* and *Mauretania* appeared only in the first decade of the 20th century when, supposedly, the time was right. Such liners were not built earlier, I suggest, because potential entrepreneurs and investors had drawn a pessimistic lesson from the saga of the *Great Eastern*: a monster ship is a high-risk venture unlikely to yield profit. An actual explanation of why the *Great Eastern* was a spectacular flop as a passenger liner is more complex.

Had the ship been placed on the eastern route, as Brunel intended and for which the ship was designed, each voyage might have carried thousands of passengers and huge cargoes. However, because the Great Ship Company was under financial duress, the directors chose the more expedient Atlantic route. The *Great Eastern* never headed east.

There was ample demand for steamship service on the North Atlantic but the *Great Eastern* was disadvantaged by postponements of its inaugural voyage, months spent on two major repairs, public relations excursions, and a charter to Canada. Closer approximation to a regular schedule and competitive fares might have allowed the *Great Eastern* to peel off business from the other 45 vessels on the route.

Another obstacle was the *Great Eastern*'s damaged reputation. Indeed, the print media were relentless in blaming the ship's many misfortunes on mystical forces: "extraordinary ill luck . . . haunts the Great Eastern" and "there never was so unfortunate a ship."[51] The branding of the ship as a jinx no doubt dissuaded many otherwise sensible people from booking passage. Ironically, the *Great Eastern* was the safest vessel on the high seas with its double hull, plentiful bulkheads, and dual-mode propulsion.

Losing money every year plying the North Atlantic and accumulating an unsustainable debt, the Great Ship Company failed and Brunel's last great technology was taken out of service. Together these several factors help us understand why the *Great Eastern*, though in many ways a splendid vessel, served only three years as a passenger liner.

Before the *Great Eastern* was sold for scrap, its fittings were auctioned. Five days of bidding realized £38,000 for the "artifacts and keepsakes of the greatest ship of the nineteenth century, the first of the superships."[52] At Anfield, home to the Liverpool Football Club, stands an old flagpole that happens to be one of the *Great Eastern*'s masts. Wood from the ship entered the Smithsonian Institution in the form of a cane fashioned for Cyrus Field.[53] In the Liverpool Maritime Museum, one can find a credenza, meat dish, coffeepot, and saloon lamp. There are also abundant "afterlife artifacts," things made in recent decades that celebrate the *Great Eastern* and cater to collectors. These include mugs, mouse pads, refrigerator magnets, postage stamps, posters, postcards, and model kits.[54]

As for the *Great Britain*, its last productive days were spent ferrying immigrants to Australia. After retirement to the Falkland Islands in 1884, "she was used as a warehouse, quarantine ship and coal hulk until scuttled in 1937."[55] In 1970 the badly deteriorated vessel was towed back to Bristol, lovingly restored, and put on exhibit. The *Great Britain* now hosts more than 150,000 visitors a year at one of the world's most impressive maritime museums.

Many of Brunel's bridges, tunnels, and other structures have survived and carry automobile and rail traffic. And the Brunel-designed Temple Meads and Paddington stations, both greatly enlarged, still serve travelers.

NOTES

[1] Unless otherwise noted, I rely on Buchanan (2001) and Vaughan (1991) for the Clifton Gorge, Great Western Railway, and *Great Western* steamship projects.

[2] Brunel quoted in Vaughan (1991:41–42).

[3] Brunel quoted in Vaughan (1991:46).

[4] Vaughan (1991:50–51).

[5] Vaughan (1991:62).

[6] Vaughan (1991:142).

[7] Buchanan (2001:200); Vaughan (1991:89).

[8] Vaughan (1991:93).

[9] For a map of the Floating Harbor and vicinity, see Buchanan (2001:47).

[10] Vaughan (1991:93).

[11] https://en.wikipedia.org/wiki/SS_Great_Western, accessed 15 January 2017.

[12] Vaughan (1991:94-95). For the time and speed on the first voyage, see Kludas (2002:146).

[13] https://en.wikipedia.org/wiki/SS_Great_Western, accessed 15 January 2017.

[14] Vaughan (1991:157–161).

[15] https://en.wikipedia.org/wiki/SS_Great_Britain, accessed 15 January 2017.

16 For technical details of the *Great Britain*, see Claxton (1845).

17 Vaughan (1991:162).

18 Vaughan (1991:162).

19 https://en.wikipedia.org/wiki/SS_Great_Britain, accessed 15 January 2017.

20 https://en.wikipedia.org/wiki/SS_Great_Britain, accessed 16 January 2017.

21 On Russell see Emmerson (1977).

22 Vaughan (1991:230).

23 https://en.wikipedia.org/wiki/Soliton, accessed 16 January 2017.

24 Vaughan (1991:230–231).

25 Vaughan (1991:231).

26 Emmerson (1977, 1981) supplied much information on the *Great Eastern*; also of use were Buchanan (2001) and Vaughan (1991). Beaver (1969) is lavishly illustrated with period images.

27 Emmerson (1977:57).

28 Emmerson (1977:74–75).

29 Emmerson (1977:83).

30 Emmerson (1977:76).

31 For a detailed description of the *Great Eastern*, including interior compartments and furnishings, see Cain (1860). On the construction process, see *Illustrated London Times*, 16 January 1858. Russell (1864–1865) contains many technical drawings.

32 Emmerson (1977:105).

33 Emmerson (1977:122).

34 "The Failure to Launch the *Great Eastern*," *New York Times*, 17 November 1957, p. 1.

35 For details on the launch, see Emmerson (1977:126–129).

36 Emmerson (1977:131–132).

37 Emmerson (1977:134).

38 "The Great Eastern: The Explosion and Its Causes," *New York Times*, 27 September 1859, p. 1.

39 "Ocean Steam Navigation," *New York Times*, 20 September 1859, p. 4.

40 Quoted in "The Great Eastern: Her Fitness for the American and Indian Trade," *New York Times*, 5 January 1860, p. 1.

41 "The Great Eastern: First Voyage of the Vessel to the United States," *New York Times*, 29 June 1860, p. 1.

42 "The Great Eastern: The Public Are Welcome at One Dollar a Head," *New York Times*, 3 July 1860, p. 8.

43 Emmerson (1981:87–92).

44 Emmerson (1981:135).

45 "The Mammoth Steamer – Great Eastern," *London Journal*, 5 January 1856, p. 296.

46 https://en.wikipedia.org/wiki/SS_Great_Eastern#First_voyage_to_North_ America, accessed 23 January 2017.

47 "The 'Great Eastern' Steam-Ship," *London Review of Politics, Society, Literature, Art, and Science*, 3 October 1863, pp. 360–361.

48 Dibner (1959); Hearn (2004).

49 Emmerson (1981:144–145).

50 Culver and Grant (1938:239).

51 "The Great Eastern," *Chicago Press and Tribune*, 10 August 1860, p. 3; "The 'Great Eastern' Steam-Ship," *London Review of Politics, Society, Literature, Art, and Science*, 3 October 1863, pp. 360–361.

52 Emmerson (1981:144).

53 Hal Wallace, curator of the Electricity Collections at the National Museum of American History, Smithsonian Institution, told me about the cane.

54 Schiffer (2013b:250–251).

55 https://en.wikipedia.org/wiki/SS_Great_Britain, accessed 22 January 2017.

6

FERDINAND DE LESSEPS'
PANAMA CANAL

On January 15, 1849, Samuel C. Upton of Philadelphia boarded the *Osceola*, a sailing ship headed to San Francisco by way of Cape Horn. In his engaging travelogue, Upham recalled his trepidation at the thought of sailing through the treacherous Straits of Magellan—notorious for its furious gales, rough seas, and icebergs.[1] Many ships had been lost there but the *Osceola* survived the passage. After 176 days at sea plus 25 days in ports along the way, Upham arrived in San Francisco and traveled to the goldfields in the mountains east of Sacramento, hoping to strike it rich. But alas he did not. Instead he founded Sacramento's first newspaper, quickly sold it, and headed back to Philadelphia.

Dreading the return voyage around Cape Horn, Upham found a much quicker route home, yet it too had perils. The trip began on the steamship *Columbus*, which deposited him and fellow passengers on the Pacific coast of Panama. There he hired two mules at $16 each—one for him, the other for his trunk. Along the way the travelers spent a night sleeping in grass hammocks and fending off fleas and mosquitoes. Next day, for travel on the Chagres River, Upham chipped in $10 to charter a native canoe that had to be bailed out during a torrential downpour. At the mouth of the Chagres River on the Caribbean coast, Upham boarded the *Falcon*, a star-crossed steamer on which seventeen passengers died en route to Havana from cholera contracted in Panama. After a brief stay in Havana, the steamship *Ohio* brought him to New York City. This trip lasted 38 days but only 32 were spent at sea. Upham never returned to California, his El Dorado.

* * * * *

Despite many dangers and discomforts, for many travelers the route through Panama was preferable to sailing around the tip of South America. Panama became especially well traveled beginning in 1849, owing to the California gold rush and the opening of California and Oregon to American settlers. To speed communication between its coasts, the U.S. government contracted with steamship lines to carry mail between the East Coast and Panama and between Panama and the West Coast. William H. Aspinwall was awarded the contract for the Pacific leg and established the Pacific Mail Steamship Company. But he had a larger project in mind: building a railroad across the Isthmus of Panama.

By 1849 railroad and steamship technologies had matured somewhat, offering capable means of transportation on land and sea. Even so, it would be difficult to build a railroad across mangrove swamps, jungle, mountains, and flood-prone rivers fed by 170 inches of annual rainfall.

Aspinwall along with Henry Chauncey and John L. Stephens, the latter a seasoned traveler in the jungles of Central America, incorporated the Panama Railroad Company in April 1849 and issued $1 million in stock.[2] The following year the company signed a contract with the government of New Granada (a country that included modern-day Colombia and Panama) to lay the rails. In addition to granting the Panama Railroad ample land and eight years to complete the work, the contract required free ports at both ends—Colón (also called Aspinwall) on the Atlantic and Panama City on the Pacific. A preliminary survey of the isthmus confirmed the railroad's feasibility and a follow-up survey by an "experienced party of engineers," led by Colonel G. W. Hughes, determined the actual route (Figure 6.1).[3]

Work began in May 1850 on the Atlantic side. The project faced serious challenges in its early going. Significantly, all materials and

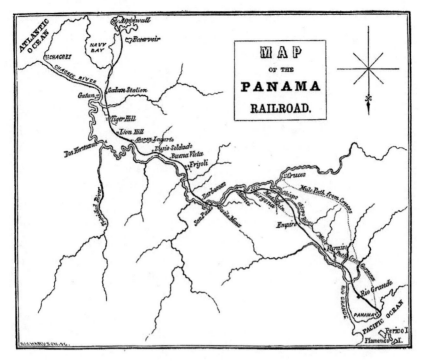

Figure 6.1. Map of the Panama Railroad.
Source: Otis (1862:v).

equipment—even wood for the ties—had to be imported. The port of Colón was soon bustling with the arrival of steamship after steamship laden with cargo destined for the project. In addition, local labor was in short supply, in part because the company was unwilling to pay a competitive wage. One solution was to import nearly a thousand Chinese men but this strategy failed, for most of them succumbed to disease or suicide. A somewhat more successful solution was importing thousands of Afro-Caribbean laborers, yet disease also took a heavy toll on these men.

In addition to penetrating swamps and jungles, the railroad had to lay tracks over the Chagres River, three hundred feet wide at Barbacoas. This required a bridge that could withstand sudden flood surges that might reach "forty feet in a single night."[4] After one flood a span was swept away and had to be rebuilt.

And finally the company had to create its own infrastructure. In addition to a hospital and employee housing, the Panama Railroad built "sidings and machinery for reversing locomotives," a spacious machine shop with "first class machinery," a blacksmith's shop with six forges, a brass foundry, facilities for making iron castings, a shop for repairing cars, and warehouses and passenger stations in Colón and Panama City.[5] Machinery was also constructed for transferring heavy cargoes such as "gold and silver ore, timber, anchors, . . . cannon shot and shells, iron-work in pieces of twenty-five tons, heavy equipment, guano, and whale-oil."[6]

Although little was accomplished in the first year, in October 1851 "a train of working cars, drawn by a locomotive, passed over the road as far as Gatun."[7] Because money from the stock offering had been spent, borrowing became necessary. Soon, however, the company figured out how to make the railroad pay for itself. As each segment of the line opened, passengers turned up, eager to avoid going even a short distance by mule or canoe.

By early 1854 the line had progressed 37 miles from Colón and work had begun on the Pacific side. According to company historian Fessenden Otis, "On the 27th day of January, 1855, at midnight, in darkness and rain, the last rail was laid, and on the following day a locomotive passed from ocean to ocean."[8] Segments of the line had been built with temporary materials and structures that had to be replaced over the next four years. When the construction account was closed at the end of 1859, the Panama Railroad had earned slightly more than the $8 million it had cost. A fortunate outcome, indeed.

In the first five years of operation, the Panama Railroad carried 196,000 passengers and almost a hundred thousand bags of mail. And travel time was reduced to around three hours. Cost of the trip: $25 for adults and $200 for a ticket that included first-class passage to San Francisco. Seven steamship lines and four lines of sailing vessels connected with the Panama Railroad. This was a very prosperous railroad and its stockholders earned handsome dividends.[9]

After mid-century railroads and steamships were helping to accelerate industrialization and foster growth of the acquisitive middle and upper classes, especially in the West. Industrialization, growing prosperity and international trade, and migration in turn intensified demand for rapid, long-distance transportation. It was, as economists say, a virtuous cycle. But many geographic barriers to speedy travel remained to be overcome by new and even more ambitious projects.

Two monumental projects that conquered such barriers were completed in 1869. The Transcontinental Railroad in the United States offered people a faster and more convenient way to travel and send mail and cargo between the coasts, and along with the telegraph made governing this vast nation more feasible. In Egypt the Suez Canal joined the Mediterranean and Red seas, enabling ship traffic between Europe, India, and the Far East to avoid the arduous route around Africa.

The creation of a water route across Egypt was an idea thousands of years old. In fact, traces of ancient canals—at least one dating to a pharaoh's reign—could still be seen when Napoléon began to explore the possibility.[10] However, the emperor's inept surveyors concluded that the two seas were at different elevations. This apparent obstacle ruled out a sea-level canal at the time, but an 1847 resurvey corrected the error and prepared the way for new canal proposals.

The prime mover behind the Suez Canal project was Ferdinand de Lesseps.[11] Retired from the French consular service, in which he had served with distinction, de Lesseps became a gentleman farmer. Although he had long dreamed about a canal to join the seas, it was during his agrarian interlude that de Lesseps began to read extensively and prepare concrete plans. Looking to the future, he expected trade between the East and West to continue accelerating; thus a "Suez Canal could develop that traffic in a marvellous [sic] manner."[12] An energetic man in his middle years, de Lesseps had a new and all-consuming ambition to benefit humanity by building the canal. The Suez project would give de Lesseps the experience—and the unbounded confidence—to later tackle the Isthmus of Panama.

Writing in 1854 to Saïd Pasha, the viceroy of Egypt and an old family friend, he opined that "the Suez Canal project is sure of the support of all enlightened persons in all countries, because it is of such importance for the future of the world, and therefore will be free from all serious opposition."[13] This cheery belief would prove to be wishful thinking. During the following fifteen years, political intrigues among many nations, the Crimean War, the ever-present threat of other wars, and the resolute opposition of Britain would require de Lesseps' utmost persistence and patience owing to tiresome negotiations and frequent travels.

Not being an engineer, de Lesseps sought scientific validation of his project. With funding from the viceroy, he assembled a party of three French engineers along with many dozens of camels loaded with water and provisions. The party "crossed the isthmus from north to south, studying the nature of the land, examining the possibility of a fresh track . . . without locks dug directly from sea to sea."[14] He opposed locks—they raise and lower ships—because they would slow ship traffic, which he forecasted might reach one hundred ships daily.[15] Several years of detailed studies, one undertaken by a commission appointed by distinguished European engineers, concluded that a lock-less canal at sea level was feasible.

In October 1858 de Lesseps announced the formation of the Compagnie Universelle du Canal Maritime de Suez, of which he was president. To raise money for the project, estimated to require £8 million, de Lesseps first approached the British-owned firm N. M. Rothschild & Sons to underwrite a stock offering. When he learned about the firm's 5 percent commission, however, he changed course, organizing the sale of four hundred thousand shares himself and establishing sales outlets in European capitals—including Amsterdam, Vienna, Paris, and London—as well as in Alexandria, Constantinople, and even New York.[16] Although confident that people would jump at the chance to own shares in his enterprise, there was very little interest in the "universal" company beyond France. And in France the company sold only 220,000 shares, the vast majority bought by small investors. The remaining shares were awarded gratis, in several large tranches, to Viceroy Saïd Pasha and his successor, Viceroy Ismail.[17]

Not surprisingly, Viceroy Saïd Pasha granted the company a concession for the project. Egypt would furnish land for both the Suez Canal itself and a second canal to bring fresh water from the Nile to the project's halfway point at Timsah. Importantly, Egypt would also supply workers through the traditional corvée system, a form of forced labor. Conforming to de Lesseps' internationalist vision, the canal would always be open "to every commercial vessel crossing from one sea to the other."[18]

The route was finalized and preliminary work begun, including testing of soils and construction of the freshwater canal. The Suez Canal would cross the desert from Port Saïd on the Mediterranean Sea to Suez on the Red Sea, covering a distance of about 170 kilometers, part of which passed through lakes.

Construction of the canal itself began on September 25, 1859, and employed up to twenty thousand corvée laborers working for contractors. "The first diggings were accomplished by *fellin* wielding shovels and short picks. . . . [S]and was removed in baskets . . . carried on men's backs or, sometimes, by mules and camels."[19] This process of excavation was nothing new. In Egypt backbreaking labor had been used for millennia to dig canals.

Although work was well along by the mid-1860s, Britain made a desperate last move to derail the project. From the beginning London opposed the Suez Canal, fearful of France's growing influence in the region. The *London Globe* was especially vitriolic, claiming that "the whole thing is a flagrant robbery gotten up to despoil the simple people who have allowed themselves to become dupes."[20] Prime Minister Palmerston and other members of the House of Lords railed against the project, more than implying that de Lesseps was a crook and the company a fraud.[21] Not limiting its opposition to heated rhetoric, London pressured Ismail—the new viceroy of Egypt—to stop supplying corvée labor. The viceroy gradually complied, but in making this move violated the terms of the concession.

Although finding some workers willing to put up with scorching heat, de Lesseps and his engineers decided on a more definitive solution to the labor shortage: the excavations would be mechanized. Machines were ordered from England, France, and Belgium, some of them designed by the company's engineers. Eventually around a hundred steam-powered machines, mainly dredges, were put into operation. Dredges might seem a curious choice for removing sand, but the company employed a novel process. A short segment of the route was flooded with water from the Nile canal. Dredges then re-

moved the slushy sand to the required depth—around eight meters—and delivered it over the bank; the process was repeated for
the next segment (Figure 6.2). Special "earth dredges" with a chain
of buckets were employed on dry segments. The first twenty
dredges arrived in 1864 and work resumed with two thousand
workers, sufficient for the tasks.[22]

With funds exhausted from the initial stock offering, the company had to raise more money. The company sold preferred shares
in 1866 on the basis of optimistic projections of ship traffic. More
significantly, Viceroy Ismail—prodded by Emperor Napoléon
III—paid £3.4 million to compensate the company for the corvée
labor withheld and agreed to complete the canal.[23]

The canal itself was navigable by fall 1868 and de Lesseps made
the first passage in fifteen hours.[24] Before commercial traffic could
use the canal, however, additional work was needed to finish the
port facilities at the two termini. Among its amenities Port Saïd

Figure 6.2. Dredge at work on the Suez Canal.
Source: courtesy of Wikimedia Commons, from Tropenmuseum, part of
the National Museum of World Cultures.

boasted a state-of-the-art lighthouse, its electric-arc lamp casting afar a pure white light. The canal was completed in 1869, in its tenth year, at a cost of about £16 million.

The ceremonial opening took place at Port Saïd on November 17, 1869, and featured a lavish fireworks display. Viceroy Ismail invited six thousand guests and paid all their expenses; also present were thousands of uninvited Europeans and, along the canal's route, many Egyptians. At the event's climax, in the imperial yacht *L'Aigle*, Princess Eugénie—de Lesseps' first cousin and the daughter of Napoléon III—led a large flotilla of ships through the canal.[25]

The *Shipping Gazette* pronounced the canal "a magnificent success" but the Suez Canal was more than a technological success.[26] In 1870, its first full year of operation, the canal served 486 ships and earned £206,373 in tolls. Just five years later, traffic had increased to 1,264 ships with tolls of £994,375—an upward trend that continued far into the future, but never reached a hundred per day.[27] Ironically, British ships were by far the canal's biggest users: in 1876, for example, 75 percent of the ships flew the Union Jack.[28] Eventually, England bought Viceroy Ismail's shares and took control of the canal it had viciously scorned.

Numerous magazine and newspaper articles in France and abroad praised the canal and lionized de Lesseps. *Scientific American* observed that "he combines a strength of will and fixity of purpose worthy of Napoleon or Caesar himself."[29] In France the enormity of his accomplishment elevated de Lesseps to the status of national hero. Among his many honors, he was inducted into the prestigious French Académie des Sciences despite lacking scientific credentials. And with Palmerston dead, England made amends. During de Lesseps' visit in 1870, he was "fêted at Liverpool, at Stafford House, at the Mansion House, at the Crystal Palace," complimented by dignitaries, awarded the Society of Arts' gold metal, and knighted by British India. He was, according to the *Spectator* magazine, "the lion of the day."[30]

✳ ✳ ✳ ✳ ✳

Ferdinand de Lesseps was 64 years old when the Suez Canal opened but he looked much younger and was still vigorous. Instead of cultivating vegetables or seeking national office, he contemplated undertaking other monumental projects such as building a railroad to Bombay and Peking (Beijing). Nothing came of these possibilities. It was a canal across the Isthmus of Panama that seized his imagination—a project worthy of his talents, ample self-confidence, and unremitting ambition. Moreover, this project was consistent with de Lesseps' earnest belief that a Panama canal would bestow benefits on humanity.

A canal across Panama had been envisioned for centuries but no company or government had committed itself to the project. Such a canal would of course have considerable value to the United States, in both commercial and military realms, but the U.S. government went no further than commissioning a preliminary survey for possible routes.[31]

Still basking in the afterglow of his successful Egyptian canal, de Lesseps leveraged that experience and celebrity to encourage international support for the Panama project. In 1879 he convened in Paris a congress of 136 engineers from European countries—primarily France—and the United States. Many canal plans were presented but de Lesseps made a strong case for a sea-level canal that mainly followed the route of the Panama Railroad. The French engineers, who perhaps had read de Lesseps' self-aggrandizing book on the Suez Canal, were in awe of what he had accomplished and trusted that he could pull off another near-miraculous canal project.[32] With only muted dissent, mainly from Americans, the congress strongly supported de Lesseps' proposal.

Buoyed by this success, de Lesseps moved quickly and formed a syndicate of wealthy investors to buy out an existing concession for a canal across the isthmus. Having obtained the coveted concession, his next move was to found the Compagnie Universelle du Canal Interocéanique de Panama in October 1880. Following his Suez model, de Lesseps snubbed the financiers and peddled the company's shares to small investors throughout France, aiming to raise F400 million (about £16 million). Remarkably, in the absence of a *detailed* geological survey of the isthmus, he assured prospective investors that the Panama Canal "will be more easily begun, finished and maintained than the Suez Canal."[33] Owing to heated opposition from French financiers, the initial public offering failed, with fewer than 10 percent of the shares sold.

In the meantime de Lesseps, his family, and an international technical commission—assigned to survey the isthmus and finalize a route for the canal—visited Panama during the dry season, traveling by train until they reached the washed-out bridge at Barbacoas on the Chagres River. The party crossed the river on foot, over planks that spanned the bridge's gaps, and boarded a train on the opposite bank. The second train proceeded to Panama City.

As historian David McCullough observes in *The Path Between the Seas: The Creation of the Panama Canal 1870–1914*, "The situation at Barbacoas should have been the clearest possible warning to de Lesseps and the others."[34] But de Lesseps believed that neither raging river nor swamps, jungles, and a mountain range would stand in his way, so sure was he that every problem had a solution. Even so, he must have suspected from the beginning that the Panama Canal would dwarf Suez in difficulty. Compared with Panama, the Egyptian desert was a salubrious environment, lacking the tropical diseases that would kill thousands of workers in the isthmus. Nearly all of the logistical and environmental problems faced by the Panama Railroad would also plague de Lesseps' project. The only saving grace was the railroad itself, which the canal company purchased at a premium in 1881.[35]

The original Paris congress concluded that building the canal would take twelve years and cost F1.07 billion (about £43 million). The technical commission, under pressure from de Lesseps, reduced the estimates to eight years and F843 million. Not satisfied, de Lesseps dropped the price still further to F658 million.[36] The latter number, he reasoned, would be more palatable to potential investors. By discounting expert opinion, de Lesseps was taking a big risk—perhaps expecting that later cash shortfalls, as at Suez, would somehow be covered. But Panama had no spendthrift viceroy.

The company was now ready to seek investors again. A new prospectus incorporating de Lesseps' unrealistic numbers was circulated. The Hero of Suez himself toured France, England, the United States, and other countries, "delivering speeches in which the enormous profits to accrue to the fortunate investors in the Panama Canal project were depicted in the seductive rhetoric that was always at his command."[37] He also bribed newspapers at home, expecting that favorable coverage of the project would influence public opinion. Apparently it did, for shares totaling F300 million were subscribed. As in the Suez project, the vast majority of subscribers were ordinary French citizens.

A new survey of the canal route was fielded and completed its work in late 1882. The findings had disquieting implications. The technical commission had originally estimated that 45 million cubic meters of material would have to be excavated. Extrapolating from the new survey information, Jules Dingler—the project's first chief engineer—expected to remove 120 million cubic meters. Although Dingler's plan required a revised timetable, the company accepted it.[38]

Work on the canal itself commenced on January 20, 1882. The completed canal was to be 74 kilometers long, 9 meters deep, and 22 meters wide at the bottom.[39] The tasks were divided among 27 contractors, including American companies. Following the Suez model,

Figure 6.3. French steam-powered bucket excavator at work on the Panama Canal, ca. 1887. Source: *Popular Science Monthly*, December 1887, p. 156.

the excavation process was mechanized with immense steam-powered excavators whose buckets were on a continuous chain (Figure 6.3). But they proved useless when reaching rock, which had to be freed by blasting and human labor. The company also used American-made steam-powered dredges and steam shovels.[40]

Workmen, mainly Afro-Caribbean but also some "negroes" from the American South, were recruited in large numbers.[41] In addition, Europeans and Americans were hired for supervisory, technical, and clerical work. Wood frame buildings with good ventilation were erected for laborers, whose numbers averaged slightly more than ten thousand over the life of the project; they earned $1.50 per day, paid in silver.[42] But turnover was very high and it was difficult to replace the many men who quit or died from disease. Estimates based on hospital records place the mortality, mainly from malaria and yellow fever, at 6–7 percent of the workforce, including Europeans and Americans (around six thousand total deaths).[43] A more realistic estimate of mortality, taking into account people who didn't die in hospitals or who left the country while ill, is about twenty thousand—including the loss to disease of Dingler's entire family.[44]

After more than two years of work, scarcely one-tenth of the excavations had been completed and money was running short. Turning to the French government, de Lesseps "applied for permission to issue lottery bonds."[45] In this scheme, which de Lesseps hoped would raise F600 million and allow the project to be completed in 1888, each bond purchaser was also entered in a lottery. Stung by widespread criticism of the project's slow pace and charges of mismanagement, and before the government could reject the lottery bond proposal, de Lesseps withdrew the application.

Three independent engineers, one sent by the government and the others hired by the company, assessed the project's progress. Their reports all converged on a grim conclusion: the project could not be completed on time and with available funds. Two reports recommended that the sea-level plan be abandoned in favor of a canal with locks. Rejecting that advice, de Lesseps went on the offensive, dragging to the isthmus "a large party of individuals, many of whom were influential in the commercial and financial circles of France."[46] This trip had the desired effect, for upon returning to

France de Lesseps received permission from the stockholders to issue a series of bonds.

Even so, conditions in Panama continued to worsen. Some contractors gave up on the project; others sued the company. Working plans were altered frequently, imparting "a general feeling of uncertainty as to the character of future operations."[47] Under severe pressure de Lesseps relented at last, drawing up plans that included "temporary" locks designed by Gustav Eiffel. The new excavation plan scaled back the cut in the mountains, which downsized the canal to the point where passage of some ships would have been precluded.

To fund the new plan, de Lesseps received government approval to issue F800 million in lottery bonds—more than his original estimate for completing the entire project. This time, however, ordinary French people failed to fall for de Lesseps' cheery promotions and less than 1 percent of the bonds were sold.

By 1888 the financial situation had become dire and many people believed that de Lesseps couldn't complete the canal. Even if the new plan had been executed successfully, the canal could not have earned enough money to continue paying dividends to stockholders and interest to bondholders—much less turn a profit. The fixed costs were simply too daunting. With bankruptcy approaching, de Lesseps' dream was coming to end.

In December 1888 the bankrupt Compagnie Universelle du Canal Interocéanique de Panama was put into receivership. Two months later, after a liquidator was appointed, de Lesseps' company ceased to exist, leaving hundreds of thousands of people holding virtually worthless stocks and bonds.[48] In 1894 the old company's assets were taken over by the Compagnie Nouvelle du Canal de Panama. With limited capital resources, the new company's excavations limped along for several years as it tried to complete the sea-level canal.[49] No doubt this goal was a major reason for widespread pessimism, on both sides of the Atlantic, about the new company's prospects. By the end of the decade, the rebooted French project was floundering.

And looming ahead was the very real prospect of an alternative canal. With the support of the public, past presidents, and powerful members of Congress, the U.S. government for many years had planned to build a canal across Nicaragua entirely under U.S. control. Yet the government, especially President Roosevelt, was willing to abandon this route if it could acquire "absolute ownership and control of the canal" across Panama.[50] The Nicaragua canal was a mortal threat to the French undertaking because it surely would have siphoned off much of the inter-sea traffic. Recognizing that its position was untenable, the new French company sold all its canal assets—including land, equipment, structures, and infrastructure—to the U.S. government for $40 million.

As is well known, the United States' Panama Canal project (1904–1914) succeeded where de Lesseps had failed, but it did follow his route and made use of lessons learned and some work already done. From the beginning, however, the American canal employed locks.[51]

A final indignity befell de Lesseps. The octogenarian was arrested in 1893, charged with bribing government officials, and put on trial. The court found de Lesseps guilty and sentenced him to a fine and imprisonment. However, the conviction was overturned because the statute of limitations had expired.

If—as many people believe—the conquest of yellow fever was a prerequisite for the success of the American canal, then it follows that the same disease must have doomed de Lesseps' project. As we have seen, however, the massive fatalities from tropical diseases were handled as an inconvenience; workers who died were replaced when possible. Despite ongoing labor shortages, there was no move to end the project on account of the staggering death toll. And the defeat of yellow fever did not end the peril for canal workers, as 5,600 people died on the American project.[52]

The largest obstacle to completing the French canal was literally a huge obstacle: a 345-foot-tall mountain that could not be avoided. To cut a four-thousand-foot-long channel in this mountain down to sea level would have required the removal and disposal of an unfathomably large volume of soil and rock. But the Culebra Cut, as it was called, suffered continuous landslides; these came as a surprise because the surveys had failed to identify the geological peculiarities that made the mountain's layers especially prone to sliding.[53] In coping with this problem, the company widened the excavations again and again, which created a gentler slope but at greater cost. By one estimate the prevention of landslides would have required the removal of a swath of mountain three-quarters of a mile wide at the top—a near impossibility. Even the United States had to mount a heroic, highly mechanized effort to widen the cut and deepen it to just 65 feet above sea level (Figure 6.4).

The de Lesseps project also failed to arrive at a definitive solution for controlling the Chagres River. The United States dammed the Chagres, creating the huge Gatun Lake that sits 85 feet above sea level and furnishes water for the locks. Ships sail across 33 miles of the lake as part of the canal's route.[54]

The project's failure was clinched by de Lesseps' loss of credibility, which resulted from more information coming to light about

Figure 6.4. Culebra Cut between the two highest hills, 1914.
Source: courtesy of the Library of Congress.

SPECTACULAR FLOPS

glacially slow progress, mismanagement of funds, and his deceitful forecasts of the canal's total cost.[55] The company spent vast amounts in Paris for interest on obligations, expenses for fund raising, and "amortization transactions." These finance-related expenses alone consumed 24.2 percent of the project's total funds; further Paris costs, such as the office along with its salaries and furniture, added another 5.1 percent. These were telling numbers, but the real scandal was de Lesseps' failure at the beginning to accept expert advice that the project would be far more expensive than he hoped. In fact, his original estimate of F658 million was just about half of what the unfinished project cost (F1.33 billion).[56] As these matters became public knowledge, de Lesseps could not expect to acquire additional funds from a public that had exceeded its limit of gullibility.

It is doubtful that more time and money could have salvaged de Lesseps' original plan. The project's fatal flaw was de Lesseps' insistence on a sea-level canal, which left the Chagres River largely untamed and the Culebra Cut incomplete. Under duress the Hero of Suez endorsed the revised plan with its "temporary" locks. However, de Lesseps could not raise the F800 million to implement the plan for the kind of canal that he had long disparaged as "inferior and therefore unacceptable."[57]

One wonders if any private enterprise in the 19th century could have built the Panama Canal, which cost the U.S. government nearly $375 million.[58] Could a company have raised this much money from stocks and bonds, especially for a project having so many obstacles and an uncertain completion date? David McCullough concludes that a Panama canal "was beyond the capacity of any purely private enterprise."[59]

✳ ✳ ✳ ✳ ✳

Egypt announced in 2013 that the building of the Suez Canal Authority in Ismailia would become a museum dedicated to displaying the canal's history.[60]

The Panama headquarters of de Lesseps' project on the Pacific side, built originally as a hotel in 1875, still stands in Panama City and is now home to the Panama Canal Museum (Museo del Canal

Interoceánico de Panamá). Exhibits include materials pertaining to the French effort.

In conjunction with the recent expansion of the canal, the Canal Authority of Panama contracted for archaeological surveys. These surveys have encountered remains of de Lesseps' canal project because vast amounts of French machinery were simply abandoned. Using sonar, a survey in Gatun Lake near Bohío Soldado discovered eighty objects—including wheels, axles, and chassis of Decauville railroad dump cars. These were raised and some set aside for conservation, presumably to be displayed later. Also found were several boilers, probably from Decauville locomotives.[61] The locations of several French graveyards are also known.

NOTES

[1] Upham (1878). Information on travel times is from pp. 217 and 380.
[2] Unless otherwise noted, information on the Panama Railroad is from Otis (1862).
[3] Otis (1862:21).
[4] Otis (1862:34).
[5] Otis (1862:40).
[6] Otis (1862:56).
[7] Otis (1862:32).
[8] Otis (1862:36).
[9] Maurer and Yu (2011).
[10] On the earlier canals, see http://www.suezcanal.gov.eg/sc.aspx?show=8, accessed 1 August 2016.
[11] An early history of the Suez Canal is Fitzgerald (1876); see also McCullough (1977).
[12] de Lesseps (1876:5).
[13] de Lesseps quoted in Beatty (1956:93); see also de Lesseps (1855).
[14] de Lesseps (1876:15–16).
[15] de Lesseps (1876:17).
[16] de Lesseps (1876:viii, 53–54).
[17] Beatty (1956:206, 212).
[18] de Lesseps (1876:viii).
[19] Calon (1994:222).
[20] *London Globe* quoted in Beatty (1956:187).

21 Beatty (1956:208).
22 Calon (1994:222–223).
23 Beatty (1956:227, 238, 249).
24 Beatty (1956:250).
25 Beatty (1956:253–254).
26 Quoted in "The Suez Canal," *New York Times*, 19 November 1869, p. 5. Fletcher (1958) discusses the effects of the Suez Canal on shipbuilding and trade patterns.
27 de Lesseps (1876:x); Beatty (1956).
28 "Suez and the Great Canal," *New York Times*, 20 April 1878, p. 3.
29 "Ferdinand de Lesseps – Chief Promoter of the Suez Canal," *Scientific American*, January 1870, p. 2.
30 "Sir Ferdinand de Lesseps," *Spectator*, 16 July 1870, pp. 856–860.
31 Unless otherwise noted, information on the Panama Canal comes from Lindsay (1912) and McCullough (1977).
32 de Lesseps (1876).
33 de Lesseps quoted in Lindsay (1912:129).
34 McCullough (1977:109).
35 McCullough (1977:135–136).
36 Lindsay (1912:136).
37 Lindsay (1912:137).
38 McCullough (1977:155).
39 Lindsay (1912:139).
40 Weld (1887).
41 Lindsay (1912:146).
42 The labor figure is the mean from 1881–1888 (Abbot 1907:101); see also Lindsay (1912:146).
43 Abbot (1907:101); Lindsay (1912:146).
44 McCullough (1977:160–161, 171).
45 Lindsay (1912:140–141).
46 Lindsay (1912:145).
47 Lindsay (1912:147).
48 McCullough (1977:202–203).
49 New Panama Canal Company (1898).
50 Abbot (1907:8).
51 McCullough (1977) is a detailed history of the American project.
52 Wong (2000).
53 McCullough (1977:166–169).
54 https://en.wikipedia.org/wiki/Gatun_Lake, accessed 11 August 2016.
55 Rodrigues (1885).

[56] Figures in this paragraph are from Lindsay (1912:151–152).

[57] McCullough (1977:195).

[58] https://en.wikipedia.org/wiki/Panama_Canal, accessed 9 August 2016.

[59] McCullough (1977: 240).

[60] http://www.dailynewsegypt.com/2013/04/08/suez-canal-authority-building-to-become-international-museum/, accessed 9 August 2016.

[61] Mendizábal (2011).

7

CRACKPOT INVENTION?

Nikola Tesla's World System

During the 1930s an elderly inventor treated members of the press to a gourmet birthday luncheon in the New Yorker Hotel, regaling reporters with tall tales of his latest discoveries and inventions. In 1935 he claimed to have measured cosmic rays traveling at fifty times the speed of light—overturning relativity theory, which he described as a "beggar wrapped in purple whom ignorant people take for a king." Two years later he predicted interplanetary communication and announced a new kind of vacuum tube that might operate at a hundred million volts, smash atoms, and produce radium for $1 a pound. More ominous was his "death-beam": silently it could deliver a burst of high-energy particles, killing millions of soldiers and destroying ten thousand airplanes 250 miles away.[1]

Not long after the 1931 debut of *Frankenstein* on the silver screen, Nikola Tesla came to personify the mad scientist-inventor who nonetheless was garnering indulgent coverage in the *New York Times*. Scientists and engineers dismissed Tesla's bizarre claims, so why did reporters pay attention to him? Apart from free food and the entertaining articles they could write, reporters appreciated that Tesla—in the words of one—was "the father of modern methods of generating and distributing electrical energy."[2] This description was not far from the mark: the technologies of alternating current (AC) that he invented are the foundation of modern life.[3]

115

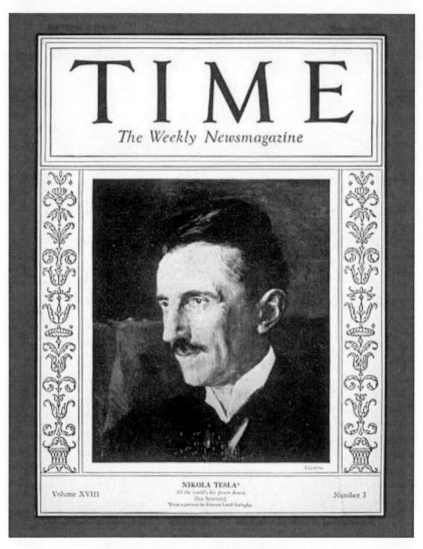

Figure 7.1. Tesla graces the cover of *Time*.
Source: *Time*, 20 July 1931.

But Tesla had done these significant works as a young man during the 1880s and 1890s; his inventions in later decades had little impact. Nonetheless, through his early accomplishments he had become a renowned electrical wizard, second only to Edison. Tesla's enduring celebrity was underscored in 1931 by his appearance on the cover and in a feature article in *Time* magazine (Figure 7.1).[4]

✳ ✳ ✳ ✳ ✳

The late 19th century witnessed a flurry of discoveries in electrical
science along with the advent of many new electrical technologies.
Among these were Wilhelm Roentgen's discovery of X-rays, Guglielmo Marconi's development of wireless (radio), Alexander Graham Bell's telephone, and Thomas Edison's light and power system.
In this heady environment, Nikola Tesla promoted his most ambitious invention yet—a "World System" for transmitting both power
and information whose vision he nurtured throughout the 1890s
and beyond. Tesla's stellar reputation enabled him to approach
wealthy New Yorkers, who furnished funds to build an experimental station in Colorado and later a lavishly equipped laboratory and
transmitting tower on Long Island, New York.

Born in Smiljan, a town in present-day Croatia, Tesla's father
Milutin was a learned and esteemed Eastern Orthodox priest; his
creative mother Djuka ran the household using some of her own
inventions. In his splendid biography *Tesla: Inventor of the Electrical Age*, W. Bernard Carlson suggests that both parents influenced
Tesla's inventive style, as he combined mental images of ideal forms
with the ability to craft real things. Layered on these early influ-
ences was Tesla's technical education at Graz and Prague in the late
1870s, when he studied electricity—especially the design of motors.[5]

Motors at that time operated on direct current (DC) supplied by
a battery or a generator. Made in countless configurations, rotary
DC motors have three major parts (Figure 7.2). First is the rotor
(or armature), which contains two or more coils of insulated wire
wound on a soft iron core. When current flows through the rotor's
coils, they become electromagnets. Second is the stationary part,
the stator, consisting of permanent magnets or electromagnets that
surround the rotor. And third is the commutator, a mechanical device that at each revolution reverses the battery's connections to the
rotor's electromagnets, changing their polarity. Interactions between the magnetic fields of the stator and rotor cause the latter to
revolve.

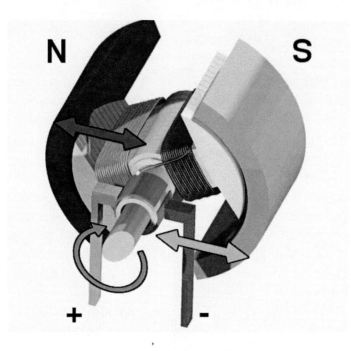

Figure 7.2. DC electric motor. Source: adapted from Wikimedia Commons, creator: Nokia123~commonswiki.

The commutator is a problematic component: its parts gradually wear out, require replacement, and create a cascade of sparks. Tesla envisioned another way to reverse the electromagnet's poles that could eliminate this necessary evil. In his embryonic vision, a revolving magnetic field would cause a rotor to turn continuously. But he had not yet figured out how to make a motor materialize this ideal.

After Prague Tesla went to Budapest, where he worked out—in his mind's eye—design features of a commutator-less motor. Instead of supplying current to the rotor's coils, he would energize the stator's many coils with AC, which would produce a rotating magnetic field. The rotating field in turn would induce a current in the rotor's coils, creating their own magnetic fields. The interactions between the magnetic fields of the stator and rotor would force the rotor to revolve.[6] In Tesla's fertile mind, this innovative design remained an ideal.

In Budapest Tesla went to work for an associate of Edison, Tivadar Puskás, who was building a telephone exchange. Tesla distinguished himself in this employment, eventually landing a job with Edison's factory in Ivry, France. There he learned more about motor and generator design, and especially how such designs were translated into actual hardware. Continuing to impress his supervisors by making improvements to generators, Tesla was sent to Strasbourg as a troubleshooter to work out the anomalies of a new Edison power station.

In his spare time at Strasbourg, Tesla sought to convert his vision of an AC motor into hardware. After many trials, in 1883 he managed to cobble together a device that confirmed his design ideal of a rotating magnetic field—but it was not yet a motor. Tesla then went to Paris where he tried to interest investors in his invention, but they were unmoved by his claim that an AC motor would revolutionize the electric power industry. Curiously, he did not apply for patents at this time, perhaps wanting to keep his invention secret until he had perfected it and could make a more compelling case to potential investors.

Fortunately, Tesla was still in the good graces of the Edison organization and in 1884—after accepting a job offer to work on generators—he sailed for New York City. In Edison's "invention factory" in Menlo Park, New Jersey, Tesla modified one of Edison's generators and also designed a complete arc-lighting system. Much to Tesla's dismay, however, the arc-lighting system was not used and he resigned his position.

After an ill-fated venture in arc lighting, Tesla was desperate for work and at last found a job digging ditches. His supervisor, who listened carefully as Tesla described his motor work, introduced him to Alfred S. Brown, a telegraph expert. Brown had collaborated with lawyer Charles F. Peck to make inroads into the monopolistic telegraph industry, since both men were familiar with how to turn ideas into commercial products. Intrigued by some of Tesla's ideas, including his design for a "thermoelectric" motor (which used heat and magnets to create motion), Brown and Peck—already wealthy from their telegraph speculations—founded a company with Tesla. In 1886 the company established a laboratory in Lower Manhattan where Tesla could perfect his inventions.

Almost a year's work on the thermoelectric generator yielded no viable design. This might have doomed the company but Brown and Peck were patient capitalists and encouraged Tesla to change directions and explore other ideas. Not surprisingly, Tesla returned to the AC motor. His starting point was a Weston dynamo, which he modified so that it could deliver several separate AC currents that were out of phase with each other (their voltage peaks came sequentially). When connected to the stator's coils by four or six wires, this "polyphase system" produced the rotating magnetic field that Tesla had long imagined (Figure 7.3). In 1888 the United States issued Tesla his first patents for AC motors and the polyphase system of distribution.[7]

At the May 1888 meeting of the American Institute of Electrical Engineers, Tesla described several versions of his motor and ex-

Figure 7.3. An early Tesla induction (AC) electric motor.
Source: courtesy of Wikimedia Commons, photograph by Stas Kozlovsky in the Nikola Tesla Museum, Belgrade, Serbia.

SPECTACULAR FLOPS

plained the polyphase system; the latter would, in principle, make possible the economical transmission of electricity over long distances.[8] Electrical engineers in attendance grasped the importance of Tesla's inventions.

Tesla was ahead of others working on AC systems in the United States and his patents were strong, but he had serious competition from European investigators who developed competent AC generators and the modern transformer (the latter was needed for a complete AC system). Able to change voltages up or down, a simple transformer has two coils—a primary and a secondary—wound on the same core of soft iron. Whether a transformer can increase or decrease the voltage depends on the ratio of turns in the two coils (or windings). For example, if ten volts is input to a primary having one hundred turns, a secondary with a thousand turns will yield a hundred volts at one-tenth the current and vice-versa.

The existence of AC generators and transformers, along with Tesla's motors and his highly efficient polyphase system, more than hinted that someone could assemble these pieces and compete with Edison. Such a move would be difficult because during the mid-1880s Edison's DC system had enjoyed hundreds of adoptions on both sides of the Atlantic, mainly in city centers and as "isolated plants" for trolleys and mansions. Business was booming.[9] Rising to the challenge was George Westinghouse, an American industrialist who had become wealthy making air brakes for railroad cars. Keen to enter the electric power industry, Westinghouse chose to go with alternating current.

Westinghouse bought up transformer, generator, and other AC-related patents but he lacked an incandescent lamp. Ironically, to compete with Edison's system he had to infringe Edison's lamp patents. Westinghouse's engineers also failed to create an AC motor and he was receptive when Brown and Peck offered to sell him Tesla's motor and polyphase patents. After much gamesmanship on both sides, Westinghouse agreed to "pay Tesla, Peck, and Brown $200,000 over a ten-year period."[10]

Edison's low-voltage DC system worked well in small-scale applications. However, current from a power station could not be sent economically for more than about a mile. (For longer distances the copper wire had to be of extreme thickness—and cost.). AC's decisive advantage was that it made possible large-scale systems. Westinghouse employed high voltage that was easily sent over tens, even hundreds, of miles and stepped it down with transformers at the consumer's end. As demand for electricity expanded beyond businesses in city centers, companies building power plants and large-scale systems necessarily adopted AC.

Casting about for new challenges, Tesla came to believe that research into high-frequency AC was an unexplored scientific frontier. During a trip to Paris in 1889, he learned about Heinrich Hertz's recent success in producing and detecting a new kind of electromagnetic radiation—invisible waves that travel through the air. Created by high-frequency AC, "Hertzian waves" would become the foundation of wireless communication (everything from radio to Wi-Fi). Tesla was captivated by Hertz's experiments, which he dutifully repeated with new apparatus.

The first fruit of Tesla's experiments, which resulted around 1890, he called an "oscillating transformer"; everyone else would call it a Tesla coil.[11] The Wizard's starting point was the induction coil, similar to one that Hertz had used in his transmitter. Invented by the American Charles Page in the late 1830s, an induction coil is a transformer having an enormous ratio of secondary to primary turns.[12] Thus the secondary's output can reach many thousands or even millions of volts. Because transformers work only on AC, experimenters prior to Tesla employed inelegant mechanical devices to convert a battery's current to AC so that it could energize the induction coil's primary.

Tesla arrived at a different solution, opting for a generator of his own design to supply alternating current. Driving the 480-coil generator at high speed, he could achieve a frequency greater than twenty thousand cycles per second.[13] When his generator was coupled to the primary of the oscillating coil, the result was a very high voltage in the secondary. Tesla also boosted the voltage by inserting a capacitor in the primary circuit.[14]

The oscillating coil produced remarkable effects new to science and technology.[15] In 1891 and 1892, Tesla exhibited these effects in lectures—first in New York City at a meeting of the American Institute of Electrical Engineers, and then in London at the Royal Institution of Great Britain, the sanctuary of science where Michael

Faraday had lectured for many decades. In a dramatic display, Tesla held in one hand "an exhausted glass tube, three feet long" and placed the other hand on a terminal of the secondary coil. Tesla was unharmed by this contact; more surprising still, the entire tube gave off a "brilliant blue light." The Wizard could also make the tube glow without any external connection. This surreal effect required two eight foot by one foot sheets of zinc insulated from their surroundings and placed about ten feet apart, each connected to one terminal of the secondary coil. Moving around in the space between the zinc sheets, Tesla held the long glass tube that "shone like a flaming sword."[16]

Reported widely, these performances gave Tesla great visibility with the general public. In describing his New York lecture, the *New York Times* promoted his reputation for electrical wizardry. "Mr. Tesla is now engaged in perfecting an improved system of electric lighting, by which, from a given amount of electrical energy, a light many times stronger than any yet produced may be secured without wires or pipes of any kind connecting with the light itself."[17] Although the press was enchanted with Tesla's demonstrations and predicted the imminent demise of the incandescent lamp, the Wizard's complicated lighting system would not be commercialized.

How did Tesla account for the seemingly magical effects of high-voltage electricity at high frequency? He believed, like most scientists at the time, that the universe was pervaded by an unseen substance called the ether. The ether was, for example, the medium through which light waves and heat propagated. Beyond this Tesla had an idiosyncratic, somewhat mystical version of the ether's properties; it could receive and transmit the universe's "infinite energy." Technologies that disturbed the ether could liberate its energy and were responsible for "electric and magnetic phenomena." His oscillating transformer, for example, strained the ether, causing waves that gave rise to the unusual effects he had exhibited—such as creating light without wires.[18]

In the conclusion to his 1891 lectures, Tesla set forth the vision that would eventually drive him to develop the World System. He believed it was possible to extract from the ether unlimited amounts of energy "not only in the form of light, but motive power and energy of any other form. . . . The time will come when this will be accomplished. . . . The mere contemplation of these magnificent possibilities expands our minds, strengthens our hopes and fills our hearts with extreme delight."[19]

Tesla enlarged upon this stirring vision in an 1893 interview published in the *New York Herald*. His plan was to use an oscillating coil of vast power to "create waves in the ether" that would set "the electricity of the earth . . . in vibration." Carefully constructed receivers, tuned to the frequency of the vibrating energy, could yield "a suitable form of power to be made available for the practical wants of life."[20] Tesla's vivid ideal was rather radical: it would be a technology that could leapfrog both AC and DC systems by placing electricity wirelessly at the disposal of ordinary people. Tesla aimed to broadcast power.

A few years later, the Wizard was honored with an invitation to speak at the Ellicott Club in Buffalo, New York, to celebrate the arrival of polyphase AC from a Westinghouse-equipped hydroelectric plant at Niagara Falls, twenty miles away. After lauding progress in applying electricity to many realms of modern life, he pointed with fatherly pride to the success of the Niagara project. This unprecedented achievement was, he said, a "glorious epoch in the history of humanity." But it was not, he emphasized, the end of electrical progress.[21]

Far from New York State, Guglielmo Marconi—a young Italian who had been tutored by physicist Augusto Righi—was using Hertzian waves to construct a system of wireless communication. Experimenting at his father's villa using an induction coil and antennas on tall poles for both transmitting and receiving, he gradually reached greater and greater distances. Satisfied that he had an invention worth developing further, Marconi offered it to the Italian government, but officials showed no interest. In 1896 he went to England where he used his mother's social connections to publicize his wireless apparatus.[22]

Government officials in London were highly receptive to Marconi's invention. In just a few years, with the help of English resources, he improved the system and formed a company to commercialize it. The Marconi Company's first successful application was ship-to-shore communication. Tesla was aware of Marconi's work and appreciated his contributions but also viewed the young man as a competitor of sorts, because the Wizard had also succeeded in communicating wirelessly with his own oscillating coil.

Telsa exhibited his mastery of wireless apparatus at the 1898 Electrical Exhibition in New York City. There he displayed in a large pool of water a small boat that he operated remotely. The transmitter was the oscillating coil with a wire antenna that could broadcast at several frequencies; receivers inside the boat had tuned circuits coupled to batteries, relays, and clockwork mechanisms. Without wires Tesla controlled the movements of his "telautomaton's" rudder and propeller.[23] This performance was a stunning display of the Wizard's inventive genius, but he denied that the telautomaton depended on Hertzian waves.[24] Rather, according to Tesla's wireless patents, his system employed electrical vibrations in the ether ("rays, waves, pulses") passing through the ground and air.[25] Marconi, meanwhile, was accumulating wireless patents based on Hertzian waves.

Having shown—as had Marconi—that information could be communicated without wires, Tesla about this time began serious work to realize his ideal of the World System. Because high voltage was necessary to transmit power, in 1898 Tesla made a machine that produced 2.5 million volts. This machine would greatly disturb the ether, causing the vast energy of the earth to flow easily and without loss through the "upper strata of the air."[26] Now, he believed, it was "quite practicable to transmit . . . electrical energy in large amounts." But confirmation of this claim would require experiments too dangerous to conduct in New York City. He would have to build an experimental station elsewhere.

In England Marconi was perfecting his system, sending messages across the English Channel and in other trials reaching distances up to one hundred miles. Unlike Tesla, who was loath to publicly demonstrate his new apparatus, Marconi delighted in promoting his system's performance characteristics. Tesla's incessant boasting about his own system—which only his assistants had seen—invited skepticism, even ridicule. *Town Topics*, a New York scandal sheet, accused him of being a "Non-Inventing Inventor" who nonetheless was receiving much press attention. Marconi, the newspaper went on to say, "telegraphs through space and Tesla talks through space."[27]

To build the experimental station, Tesla needed money—lots of money. He managed to interest John Jacob Astor IV, heir to a large fortune and a techno-enthusiast, in his wireless lighting system. With $30,000 from Astor and money from other investors, Tesla set up shop in Colorado Springs, Colorado. There he could safely conduct experiments with very high voltages. Instead of working on lighting, as Astor expected, Tesla built a system to broadcast power.[28] On Tesla's arrival in Colorado Springs in May 1899, a reporter prodded him to reveal his plans. Telsa replied, "I propose to send a message from Pike's Peak to Paris."[29] Achieving this transmission would put him decisively ahead of Marconi.

In a matter of months, Tesla's experimental station took shape in a large barn-like wooden building equipped with an enormous oscillating coil (the transmitter) and associated components. A tall pole with a metal sphere at the tip projected skyward through the roof. With a Westinghouse transformer, Tesla raised the local power to 20,000–40,000 volts and fed it into the oscillating coil's primary (now called the "magnifying transmitter").[30]

In Tesla's latest thinking, the earth would be the principal conductor of electricity with the atmosphere completing the circuit. His sensitive measurements of the changing voltage between the atmosphere and the earth during long-lasting lightning storms suggested that he could now "transmit both messages and power around the world."[31]

When operating at full power with tens of millions of volts flowing from the secondary, the magnifying transmitter created frightening effects. Outside the laboratory Tesla witnessed "full-fledged bolts of lightning darting into the air, accompanied by a barrage of tremendous crashes of thunder."[32] Some lightning bolts extended more than one hundred feet from the metal sphere. Inside the laboratory the magnifying transmitter generated huge sparks—streamers—more than twelve feet long.

Tesla took his sensitive receiver outside the laboratory and detected electricity at increasing distances. Although his observations went no farther than a mile, Tesla calculated that he could send power more than a hundred miles. As Carlson points out, "[O]nce he had detected sparks a mile away from his transmitter, Tesla had all the evidence *he* needed to be convinced that his system worked."[33]

To convince others of the great power and potential of his complex apparatus, Tesla employed a different kind of evidence: photographs of the magnifying transmitter at work. Tesla invited Robert Underwood Johnson, editor of the *Century Magazine* and a longtime admirer, to send a photographer to Colorado. Johnson dispatched Dickenson Alley, a noted art photographer who used a large-format camera to create 68 dramatic images.[34] One iconic image shows many long streamers with Tesla seated nearby, calmly reading (Figure 7.4). This photograph was a double exposure, what used to be called "trick photography."

Figure 7.4. Tesla's Colorado Springs laboratory. Source: courtesy of Wikimedia Commons, Dickenson V. Alley, photographer.

Pictures in hand, Tesla returned to New York in January 1900 and began courting investors to help him take the World System to the next stage. The images, however impressive, did not show that he could actually broadcast power and information far afield. To make this case, Tesla had to rely on his reputation, his social network in New York City, and his exotic charm and persuasive powers.

One of his first moves was to write an article for the *Century Magazine*, which not only included nine photographs from the Colorado shoot but also presented exaggerated claims of his achievements there—such as demonstrating reception "up to a limited distance of about six hundred miles." His discoveries, he claimed, "opened up the wonderful prospect of transmitting large amounts of electrical energy for industrial purposes to great distances without wires."[35] Broadcasting power this way, Tesla insisted, would be economical. As for communication, his calculations showed that messages could eventually be exchanged with Mars.[36] In the meantime he was confident that he could succeed in "effecting transoceanic communication."[37] Scientists and electrical engineers criticized Tesla for his unbridled speculations, especially after he claimed to have received signals from Mars.

Tesla talked up his plans for the World System to potential investors and attracted the interest of J. P. Morgan, America's foremost capitalist. The appeal to Morgan centered on how his wireless system could eventually replace undersea telegraph cables and more. Morgan agreed to loan Tesla $150,000.[38] The Wizard also obtained a loan from Westinghouse, provided that Tesla buy Westinghouse equipment—and he did.

Now that Tesla had the resources to continue developing the World System, he had to find a place to erect a new laboratory and transmitter. At Shoreham, on the eastern end of Long Island, he acquired a two-hundred-acre parcel from James S. Warden. Easily reached by train from New York City, the property was called Wardenclyffe.[39] Tesla also obtained the services of his friend, famed architect Stanford White, to design a large one-story brick laboratory; construction began in September 1901. Behind the laboratory the transmitting tower—a wooden derrick to be topped by a hemispherical steel frame sheathed in copper—was also being built. The entire mushroom-like structure reached a height of 185 feet (Figure 7.5).

Scarcely two months after construction began, Tesla received bad news from abroad. Marconi was in Newfoundland trying to stimulate sales of his wireless apparatus, when on December 12 he repeatedly received the letter "S" in Morse code sent from his powerful transmitter in Cornwall, England. Marconi, not Tesla, had been first to bridge the Atlantic wirelessly. Owing to widespread newspaper coverage of this feat, including the front page of the

Figure 7.5. Tesla's Wardenclyffe laboratory and transmitting tower. Source: Reeve (1911:97).

New York Times, progress in wireless communication was now firmly associated with Marconi.[40]

Tesla put on a brave front, belittling Marconi's achievement and stating in an interview that sending short messages was not nearly as important as transmitting power.[41] Writing to Morgan, Tesla also added that Marconi's apparatus—which had been designed with the help of electrical engineer J. Ambrose Fleming—employed essentially the same circuit and components as Tesla's, including a Tesla coil.[42] After all, the Wizard's earlier wireless patents specified the kinds of hardware required.

To boost interest in the World System, Tesla published a brochure that described his apparatus and communication goals. His claims, as one early biographer put it, "seemed fantastic."[43] The long list of promises included interconnection of all telegraph and telephone exchanges worldwide; "universal distribution of general news, by telegraph or telephone"; distribution of facsimiles of written materials and reproduction of photographs; "establishment of a world system of musical distribution"; and a system to enable ships anywhere to navigate without compasses.[44] What did the World System offer to an ordinary person? According to Tesla, "[A] simple and inexpensive device, readily carried about, will enable one to receive on land or sea the principal news, to hear a speech, a lecture, a song or play of a musical instrument, conveyed from any other region of the globe."[45] No wonder Tesla has been repeatedly called a man far ahead of his time. Curiously, the brochure did not mention the distribution of power.

Attempting to bring his exciting visions closer to reality, Tesla spent extravagantly on his laboratory, transmitting tower, and equipment. Soon he needed an infusion of capital. In the midst of the financial panic of 1903, Tesla wrote to Morgan, begging and cajoling him to increase his investment in the project. The financier was preoccupied with more pressing matters, such as keeping the federal government from dismantling his industrial empire. Morgan firmly refused Tesla's repeated entreaties.

Although Tesla's pitch to potential investors emphasized the World System's communication capabilities, he was still intent on broadcasting power. Writing in the *Electrical World and Engineer*, Tesla claimed he had designed a plant that would transmit at a hundred million volts—its power reaching around the globe, able to illuminate homes and furnish power to clocks.[46]

Despite the continuing cash crunch, the laboratory building was done, its equipment in place. Even though the dome's copper sheathing was incomplete (Figure 7.5), the transmitter could still operate; by mid-1903 Tesla was testing the system. Perhaps venting frustration over Morgan's rejection, on one occasion Tesla fired up the transmitter and sent lighting bolts thundering into the night sky.[47] Attracted by accounts of the curious nocturnal happenings, reporters showed up at Wardenclyffe but the Wizard did not explain what he was doing.

Tesla carried on for several more years, testing and tweaking the equipment, but his money problems worsened—as did his mental state. In 1905 Tesla suffered a nervous breakdown, dismissed the laboratory staff, and abandoned Wardenclyffe for good. There would be no World System.

✳ ✳ ✳ ✳ ✳

The cause of the World System's failure is more fundamental than Tesla's inability to raise enough money to sustain the project. According to Carlson, "Tesla believed that what he imagined about transmitting power through the earth must be true" because, as an ideal, it was the product of his mind.[48] Carlson arrives at a stark conclusion about the imperfect relationship between Tesla's ideal and reality: during several years of testing the apparatus, Tesla learned that it couldn't send power to great distances. "Hence, when Tesla could not get Wardenclyffe to work the way he wanted, he must have been confronted with a serious dilemma: either he was wrong or nature was wrong. Unable to accept either alternative, Tesla suffered a nervous breakdown."[49] Carlson comes down on the side of the physicists who judge that Tesla's principles about how electricity behaved in the earth were mistaken. Broadcasting power in the way Tesla proposed was simply impossible.[50] There was, after all, no ether.

Yet because he was transmitting at high frequencies, Tesla could have sent messages much as he had envisioned. After all, high frequencies create Hertzian (electromagnetic) waves that can travel around the world. But Tesla was preoccupied with sending power, not signals. While Tesla fiddled with his apparatus, Marconi in England, Reginald Fessenden and Lee de Forest in America, and others in Europe were using Tesla's inventions and circuit designs to establish wireless communication as a commercial technology.

Undeniably a brilliant thinker and inventor, Tesla did not bother with the details of commercializing an invention—not wireless communication, not anything. In fact, many of his minor inventions, had they been developed further, manufactured, and brought to market, might have earned him money.[51] But the Wizard didn't care to make and sell products. When his assistants set up an assembly line in the laboratory to make small Tesla coils for sales to schools and experimenters, an aloof Tesla left them alone.[52]

Tesla believed that financial men should support his projects because he was Nikola Tesla, the genius who had given AC power to the world. He did not aim to build an industrial empire founded on his inventions, for he preferred to spend all his time inventing. The Wizard would create ideals in his mind, develop principles, do experiments, and write patent applications. If the applications were successful, Tesla might license or sell the patents. So focused on his laboratory work was Tesla that he rarely sued other inventors and manufacturers for patent infringement. One conspicuous exception was Marconi.

Tesla's case against Marconi hinged on who had invented the basic principles of wireless communication. Lawsuits involving Tesla, Marconi, and the U.S. government dragged on for decades. In 1935 the Court of Claims invalidated Marconi's fundamental patent, arguing that its claims had been anticipated in the patents of Tesla and other radio pioneers. The U.S. Supreme Court affirmed this ruling in 1943, the year Tesla died.[53]

No traces remain of Tesla's Colorado Springs experimental station. Wardenclyffe fared somewhat better. In 1917 the transmitting tower was salvaged for materials but its foundation remains. And the laboratory building, which hosted a series of post-Tesla industrial users, still stands. A sixteen-acre tract that encompasses the entire experimental station is now owned by the Tesla Science Center at Wardenclyffe, which hosts a variety of fundraising events. A museum and education center are planned.[54]

NOTES

[1] On cosmic rays see "Tesla, 79, Promises to Transmit Force," *New York Times*, 11 July 1935, p. 23; "Sending of Messages to Planets Predicted by Dr. Tesla, on Birthday," *New York Times*, 11 July 1937, p. 29; "Tesla, at 78, Bares New 'Death-Beam,'" *New York Times*, 11 July 1934, p. 18.

[2] "Sending of Messages to Planets Predicted by Dr. Tesla, on Birthday," *New York Times*, 11 July 1937, p. 29.

[3] On the adoption of AC systems, see Hughes (1983).

[4] *Time*, 20 July 1931, pp. 29–31.

[5] Unless otherwise noted, information in this chapter is from Carlson (2013). Several websites also compile, and link to, many primary and secondary sources. Especially useful: http://www.tfcbooks.com/default.htm, accessed 22 June 2016.

[6] Carlson (2013:52).

[7] On the early history of the AC or "induction" motor, see Kline (1987).

[8] Tesla (1888).

[9] On Edison and his works, see Israel (1998).

[10] Carlson (2013:113).

[11] On the Tesla coil, see Carlson (2013:117–125).

[12] Page (1867).

[13] "Tesla at the Royal Institution of Great Britain," *Scientific American,* March 1892, p. 168. He claimed that it was possible in principle to achieve sixty thousand cycles per second from a similar machine (Tesla 1891a).

[14] Tesla (1891b:312–314).

[15] Tesla (1891c).

[16] Tesla (1892:168).

[17] "Wireless Electric Lamps," *New York Times*, 9 July 1891, p. 4.

[18] Tesla (1891b:267, 268).

[19] Tesla (1891b:318-319).

20 "Scientists Honor Nikola Tesla," *New York Herald*, 23 April 1893, p. 31.

21 Tesla (1897:47).

22 On the development of wireless communication, see Aitken (1976).

23 Tesla U.S. patent #613809. Tesla (1900:184–187).

24 Tesla (1900:197).

25 Tesla U.S. patent #613809.

26 Tesla (1898:345).

27 *Town Topics* quoted in Carlson (2013:260–261).

28 On Tesla's Colorado sojourn, see Carlson (2013:301).

29 Tesla, quoted in Carlson (2013:265).

30 Carlson (2013:267–269).

31 Carlson (2013:273).

32 O'Neill (1944:186).

33 Carlson (2013:295), emphasis in original.

34 Carlson (2013:295–300).

35 Tesla (1900:209).

36 Carlson (2013:265).

37 Tesla (1900:209).

38 Carlson (2013:311–318).

39 Carlson (2013:319).

40 "Wireless Signals Across the Atlantic," *New York Times*, 15 December 1901, p. 1.

41 Carlson (2013:334).

42 Carlson (2013:337–338).

43 O'Neill (1944:209); the brochure is on pp. 209–211.

44 O'Neill (1944:210–211).

45 Tesla (1905:23).

46 Tesla (1904:430).

47 "Tesla's Flashes Startling," *New York Sun*, 17 July 1903, p. 1.

48 Carlson (2013:364).

49 Carlson (2013:365).

50 Tesla was of course broadcasting power as electromagnetic radiation, but its strength fell off so greatly with distance that his ideal was out of reach.

51 O'Neill (1944:213).

52 Cheney (1981:167–168).

53 Carlson (2013:377–378).

54 http://www.teslasciencecenter.org/, accessed 16 December 2017.

8

VISIONARY INVENTOR

R. Buckminster Fuller's
Dymaxion World

Inventor, philosopher, and futurist, R. Buckminster Fuller is best known for developing the geodesic dome. He was also a prolific lecturer and author, propounding a philosophy of sustainability and resource conservation that still resonates today. Believing that the technologies of everyday life wasted materials and labor, Fuller designed radical alternatives. Although his designs for automobiles and houses didn't upend entrenched industries as he hoped, they did generate publicity and helped establish his reputation as a technological visionary. Fuller's immense creativity was buttressed by his belief, unshaken by the horrific events of mid-century, that technology can solve any problem.[1]

Throughout his long life, Fuller offered extraordinary visions of future technologies—such as a dome covering part of Manhattan, an underwater city, and a city floating in the clouds—each seemingly more outlandish than the last, echoing and influencing science fiction. His only remarkable success was the geodesic dome, which followed several spectacular flops.

✳ ✳ ✳ ✳ ✳

Although born into a family of comfortable means, Fuller's early life was not auspicious. Expelled from Harvard twice and never earning a college degree, he became a machinist in Canada. This was a formative period, for he "came to know shop foremen, molders, machinists and their respective tools.... Sometimes I succeeded in designing better parts."[2] During World War I, Fuller served in the U.S. Navy. After the war a failed venture with his architect father-in-law, excessive drinking, and the death of a child provoked an existential crisis and thoughts of suicide. But in 1927 a dissipated and directionless Fuller had a life-altering epiphany. He committed "to discover what, if anything, an individual human eschewing politics and money-making can do effectively on behalf of all humanity."[3]

During the next few years, Fuller's philosophy took shape as he grappled with central problems of human existence.[4] He believed that planet Earth provided sufficient resources to sustain even a rapidly growing humanity, but that these resources were being squandered by antiquated and inefficient technologies. It was possible, he insisted, to "do more with less," a design imperative repeated throughout his career.

Fuller fancied that a small number of invariant laws of the universe could be applied to invent modern, efficient technologies. He aimed to identify and apply these scientific laws for the betterment of all people. He invoked the authority of science as a rhetorical device in ponderous prose and long-winded lectures. Captivated by the triangle shape, Fuller insisted that it was "scientifically right" as a basic unit of construction.[5] This was a curious claim with no foundation in physics or engineering. The triangle was a design choice dictated not by scientific law but by personal preference.

Fuller's first target was the home construction industry. In many parts of the United States, middle- and working-class houses were built mainly of wood. Each board was sawed by hand and hammered into place. The interior studs framed the walls and ceiling and were covered by lath (wood slats) and plaster; the exterior was finished with stucco or wood siding. The roof consisted of wood beams covered with wood shingles or laid with tarpaper. Plumbing of kitchen and bathroom facilities required a tangle of hidden pipes, and electricity a tangle of hidden wires. Materials for each house were purchased from local lumberyards and hardware stores. Erecting the house employed a host of specialized workers and took many months. In Fuller's view the "modern" house of the late 1920s was based on an atrocious design perpetuated by uncreative architects and a hidebound construction industry.

Why, Fuller wondered, couldn't prefabricated houses be mass-produced in factories, flown by airship to the construction site, and quickly assembled by a small team? The idea of prefabricated homes, though much disparaged by architects, had been around for decades. Since 1908 Sears, Roebuck & Company had been successfully selling mail order house kits, in many styles and prices, through its catalogs.[6] As a centralized source of house components when railroads still reached most small towns, the Sears plan achieved some gains in efficiency. However, the Craftsman houses themselves were made of traditional materials and conformed to traditional designs.

In designing his "machine-for-living," Fuller proceeded as if he knew nothing about conventional house designs, preferring to work entirely outside the existing technological traditions. Thus most features of his house lacked precedents. He even rejected the use of traditional materials: wood and plaster were out, aluminum and several kinds of plastics were in.[7]

Fuller debuted his house design in 1929 accompanied by imaginative drawings and a model several feet in diameter (Figure 8.1).[8] During public promotions, which attracted some financial support, audiences learned that the house was stunningly original in every respect, even in name. Fuller initially called it the 4D house, based on the title of his first book (1928), an allusion to Einstein's four-dimensional space-time. In preparing to exhibit Fuller's model, the Marshall Field department store in Chicago assigned Waldo Warren—an advertising specialist—to craft a catchier name. Warren combined the words *dynamic*, *maximum*, and *ion* and created "dymaxion" (dye-max'eon).[9] With Fuller's approval the 4D house became the Dymaxion house. The inventor also adopted Dymaxion as his brand and trademark and applied it to his designs. He became the Dymaxion Man creating a Dymaxion World.

Figure 8.1. Fuller and the model of the 1929 Dymaxion house.
Source: by permission of the Estate of R. Buckminster Fuller.

Fuller envisioned that the house's components could be delivered even to the remotest region and quickly assembled; it would be the universal house. To approach universality the house needed to be off the grid and largely self-contained. A diesel generator produced electricity, an artesian well or delivery truck furnished water, and a septic tank and incinerator handled waste.[10]

Roofed with aluminum sheets, the five-room house was a hexagon, the space divided into triangular rooms (two triangles formed the living room). Surprisingly, the living floor was nine feet above ground level but an elevator—no stairs—was included. More surprising still, the ceiling and floor were suspended by piano wires from a central column. The spinal column, whose core was made of aluminum tubes, also supported the utilities and eliminated the usual tangle of long wires and pipes. In the ample space below the floor, the owner could park a car or airplane. The floor, by the way, was made of a soft, inflatable material. "It will be perfectly safe for baby to fall on Dymaxion floors. He may bounce, but he will never break."[11]

Fuller emphasized that the house served not only to "keep its occupants warm and dry but [also] to reduce the drudgery of cleaning and cooking and washing and service to the minimum now permitted by the state of the mechanical arts."[12] To that end the house included an automatic washing machine, dishwasher, and vacuum cleaner.

The walls were triangular sheets of plastic (opaque, translucent, or transparent) having dual panes for insulation. The modular bathroom—including floors, walls, sink, and bathtub—was cast as one piece and mated to the water and drain pipes in the central column. In other rooms plastic furniture was poured along with the walls. An exception was the pneumatic bed, an air mattress on which adults could bounce but not break. The unusual lighting consisted of "a series of prisms and reflections spread[ing] the light of a single bulb evenly over the entire ceiling through which it delivers a diffused glow into the rooms."[13] The air was maintained at a comfortable temperature, permitting the residents to go about their Dymaxion lives in the nude.

Fuller estimated that the house would weigh only six thousand pounds and sell for no more than $3,000. He also stated that it could withstand a one-thousand-mile-per-hour wind, survive a flood or fire, and be burglar proof.[14] Although such claims left some observers believing he was a crackpot or megalomaniac, others bought into Fuller's imagined future.

Adults in the late 1920s and early 1930s, who had witnessed the birth of the automobile as well as the seemingly magical performances of radios and airplanes, were accustomed to the advent of novel technologies and might have believed Fuller's house to be somewhat plausible. After all, the workings of the Dymaxion house were comprehensible and descriptions and images of the model were compelling. Meticulously reporting a Fuller show-and-tell, an enthusiastic journalist was convinced that "the advantages he claims for the proposed Dymaxion residence are important enough to alter the entire course of the human race."[15]

Famed poet and author Archibald MacLeish also wrote effusively about the house. "[I]t may well be the prototype of a new domestic architecture. . . . [I]t will destroy the great architectural dogma that a house is what our great-grandfathers called a house, and that the architect's sole opportunity is to modify what already exists."[16] In the *Washington Post*, Frances F. Zala observed that "the occupants of the house will be free of daily drudgery, free to study, to loaf, to play, to develop the million interests that change life to living. . . . [It will] revolutionize daily living for all races and nations."[17]

Building a prototype of the Dymaxion house would have been costly, in part because Fuller's design lacked detailed specifications and incorporated expensive materials. No wealthy investors or companies accepted the challenge of perfecting and materializing the house. Decades later Fuller explained the failure by claiming that the house was not "then ready for industrial production and distribution" because it depended on foreseeable—but not yet realized—new materials such as high-strength aluminum and very stable plastics.[18] This partial explanation omitted two important considerations. First, from an engineering perspective the house's design involved major uncertainties that Fuller had glossed over, such as whether a comfortable temperature could be maintained in all environments and whether the model could be effectively scaled up. Second, during the Great Depression potential investors may have been wary of placing bets on a complex, undeveloped technology with an uncertain market. A full-scale Dymaxion house would remain in the conceptual realm for some time.

In the meantime Fuller turned his attention to the Dymaxion car, throwing down the gauntlet to the automobile industry. His original plan was even more ambitious. He would use the car to test design features leading to construction of an "omni-medium transport vehicle" capable of driving, floating, and flying.[19] Existing cars, Fuller observed, were heavy and not aerodynamic, which resulted in a lavish use of materials and—at higher speeds—poor gas mileage. By testing models in wind tunnels, aeronautical engineers had already shown that a teardrop shape reduced wind resistance. Their findings would soon be applied to streamlining passenger aircraft and Fuller used them to streamline his car.

By the early 1930s, when Fuller began building the Dymaxion car, streamlining was well on its way to becoming a design fad for both consumer and industrial products. As art historian Donald J. Bush remarks, "[T]he streamlined form provided a clear and optimistic symbol of progress to Americans struggling with the confusion and chaos of the economic Depression of the 1930's."[20] Throughout the thirties streamlining went mainstream, altering the appearance of everything from pencil sharpeners to radios to locomotives—even some cars. But Fuller didn't follow style fads or fashions. For him streamlining an automobile served one purpose only, and that was to improve fuel economy.

To build the car, Fuller formed the Dymaxion Corporation and partnered with William Starling Burgess, an aeronautical engineer and marine architect of considerable accomplishment. And he tapped sculptor Isamu Noguchi to help prepare sketches. Philip Pearson, a wealthy Philadelphian captivated by Fuller's vision, helped fund the project—as did an inheritance from the inventor's mother.

Renting space in the dormant Locomobile factory in Bridgeport, Connecticut, Fuller hired a crew of highly skilled men to work under Burgess' direction. In the next year or so, the Fuller–Burgess team built three prototype Dymaxion cars.[21] Adhering to his earlier iconoclasm, Fuller and his team worked largely outside the existing technological traditions of the automobile industry. Not surprisingly, when the three-wheel car made its public debut in 1933, it resembled nothing else on the road. Aerodynamics had the highest priority in determining the vehicle's shape, which was so unusual that journalists struggled to find words to describe it. If whales had wheels . . .

Inside and out the Dymaxion car departed from convention, including a unique drive train and body. Virtually every car built in the early 1930s had a front engine that drove a pair of rear wheels. Not so the Dymaxion. Its two front wheels were powered by a stock Ford V-8 engine of less than one hundred horsepower placed near the rear. The car was steered by a lone back wheel. The Dymaxion's body was aluminum (not steel) and the sheet metal pieces were attached to an ash skeleton, reflecting Burgess' yacht-building expertise. Fuller's choice of the more expensive metal—as well as the canvas roof in Model 1—was again dictated by the need to limit the car's weight, though the woodwork must have piled on many extra pounds. In addition to thin shatterproof glass, including a half-dozen triangular pieces in the windshields of Models 2 and 3, "the car featured air nostrils, air-conditioning, and rear view periscopes for both front and back seats."[22]

Some information is available on Model 1's performance characteristics. Owing to Fuller's embellishments and inconsistencies, however, there is ample reason to be skeptical of his claims. At around nineteen feet long, the car was said to seat eleven people, a number that perhaps included small children. On its first test run in Bridgeport, three thousand people witnessed the car traveling at 70 miles per hour, but Fuller claimed that it could reach 120 miles per hour.[23] At this speed, however, the car would have risen from the road, almost taking flight, but without the inflatable wings that Fuller projected for his omni-medium vehicle. At very low speeds, the car was highly maneuverable, for the rear wheel allowed it to turn in a very tight circle.[24] But the rear wheel also caused stability problems at higher speeds, especially in windy conditions—a problem that was only partially solved in progressively heavier Models 2 and 3.

Hinting that the car was most fuel-efficient at one hundred miles per hour, Fuller asserted that it got forty miles to the gallon but he did not specify actual test conditions.[25] Taken at face value, this number implies that the Dymaxion car was more than twice as fuel-efficient at high speeds as a 1933 Ford with the same engine. Perhaps it was. In any event, there is a dearth of reliable data on the Dymaxion's actual gas mileage.

While carrying an English aviation expert to the Chicago airport, Model 1 was struck by another car being driven at high speed by an influential city official. Both cars rolled over and the Dymaxion's driver was killed. By the time reporters arrived, the other car had been towed away, leaving the impression that Fuller's car had flipped on its own.[26] Not surprisingly, newspapers erroneously described a one-car crash. Despite this disaster, Fuller moved ahead, repairing and selling Model 1. Even though money was tight, Fuller's team managed to build Models 2 and 3, which had rear dorsal fins.

To celebrate its centennial, the city of Chicago hosted a world's fair in 1933–1934 whose theme ("A Century of Progress") was technological innovation. Not surprisingly, the Fuller car was exhibited but not alongside other cars. Sometimes Model 2 or Model 3 could be seen at the fair parked in front of architect George Keck's futuristic Crystal House (Figure 8.2). The odd-looking house and odd-looking car attracted much attention and largely favorable coverage in newspapers and newsreels. After its duty at the fair, the car was sold to Evangeline Stokowski, wife of orchestra conductor Leopold Stokowski. To pay off debts to his mechanics, Fuller gave them the remaining Dymaxion car.[27]

These few sales indicate that even deep in the Depression some people would buy novelty cars—and the Dymaxion was little more than that. To convert Fuller's brainchild from a novelty into a mass-produced product would have required a huge infusion of cash and a thorough redesign informed by conventional practice. Although Fuller received feelers from several car companies, no deal was struck. In addition to the difficulties of redesigning the car and tooling up for mass production, carmakers faced a more disturbing obstacle: the car's design was so revolutionary that it threatened an industry dependent on annual incremental changes in style.[28]

Although Pearson and Burgess wanted to continue the project and eventually put the car into production, Fuller shut it down. There would be no omni-medium vehicle. He plunged into other projects, his fertile mind churning out numerous design ideas—including the prefabricated Dymaxion bathroom, which went nowhere.[29] In later years Fuller bristled when others mentioned the Dymaxion car's "failure." He insisted that the project "was a resounding success in proving my principles and teaching me what I wanted to learn from it."[30] But this claim doesn't explain why Fuller suspended development of the car's flying and floating capabilities. Perhaps what he really learned was that his design for an omni-medium vehicle was unworkable and unlikely to find new backers.

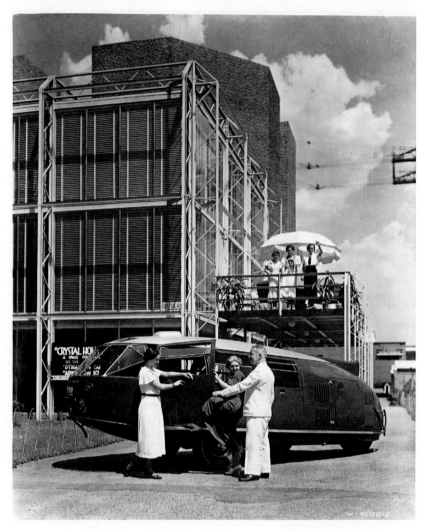

Figure 8.2. Fuller shows off a Dymaxion car at the Chicago World's Fair (ca. 1934). Source: by permission of the Estate of R. Buckminster Fuller.

* * * * *

In 1940, with U.S. involvement in the war seemingly imminent, Fuller developed the "Dymaxion deployment unit." This was a spare round structure lacking the complexities and amenities of the original Dymaxion house, yet suitable for soldiers. Following

Fuller's design, which he had adapted from a cylindrical corn bin, Butler Brothers of Kansas City built between one hundred and two hundred of them for the Army Signal Corps (Figure 8.3). The steel structure consisted of one large room with furniture and appliances arrayed around the wall. The Signal Corps' Dymaxion deployment units were successfully deployed, even in North Africa and the Arctic, but production ended when supplies of steel tightened during the war.[31]

In wartime Fuller held several positions in the U.S. government. From that vantage point, he and many others predicted that the war's end would create a conversion problem for the aircraft industry. Accustomed to churning out tens of thousands of planes annually (96,318 in 1944 alone), the companies would have to find other uses for their factory buildings, equipment, and expertise in aluminum fabrication.[32] Otherwise firms would close and tens of thousands of manufacturing jobs would be lost. It was also easy to

Figure 8.3. A Dymaxion deployment unit (1941).
Source: courtesy of the Library of Congress.

predict that millions of GIs coming home after the war would marry, start families, and create an unprecedented housing shortage. In these predictions Fuller saw an opportunity to revive and revise his plan for the Dymaxion house, which the aircraft industry could mass-produce. According to Fuller, there was "no basic difference between the fabricating of aluminum parts for the Dymaxion house and for the fuselages of B29s."[33]

Labor unions were enthusiastic about the possibility of building houses in aircraft factories. Even before the war ended, officials of the International Association of Machinists introduced Fuller to Jack Gaty, general manager of Beech Aircraft in Wichita, Kansas. After listening to Fuller present his ideas at length, Gaty evinced mild interest. Although unwilling to underwrite the project, Gaty offered to rent factory space to Fuller and provided—on a reimbursement basis—access to engineers and supplies. Agreeing to these terms, Fuller and a group of backers founded Dymaxion Dwelling Machine. The inventor resigned from the government, relocated to Wichita, and designed a new version of the Dymaxion house.[34]

After learning about the house project, the U.S. Army Air Corps contracted with Fuller's company in 1944 to build two prototypes. The first prototype, called Barwise, was completed a year later; it was cylindrical and had slightly more than one thousand square feet of interior space. Among the many design ideas Fuller drew from his first house was the use of steel wires to suspend the ceiling and floor from a central column, which hosted the utilities. The house included a refrigerator, dishwasher, and washing machine. Aluminum (for roof and walls) and plastic (Plexiglass for windows) were used extensively. There were also new features: bedroom closets containing motorized, revolving shelves; a plywood floor resting on aluminum beams; and a stainless steel fireplace. Although parts for two prototypes were manufactured—no part allegedly weighed more than ten pounds—only Barwise was built.[35]

Not long after Barwise was erected in the factory, *Fortune* published a well-illustrated article about Fuller that featured the Dymaxion house and appeared on the magazine's cover. This dwelling machine, wrote the author, "is likely to produce greater social consequences than the introduction of the automobile." A federal official who had seen the house first hand proclaimed it "the house of the future." It would, he continued, "precipitate an industrial revolution or [be] the most monumental flop in history."[36]

Some inventors tinker endlessly with their designs, taxing the patience of backers eager to commercialize the product. That Fuller was neither a trained architect nor an engineer may have contributed to his interminable revisions, for he never completed a production-ready design. However, let's give Fuller the benefit of the doubt: he was in the best position to know that the house needed more work.[37] Wielding veto power, he rejected his backers' plan for raising the estimated $10 million to produce twenty thousand houses annually. Fuller's opposition to the plan—in the face of 3,500 unsolicited orders that had his backers salivating—created an impasse but he did not relent.[38] Lacking a final design and the resources to proceed, the company folded in 1946. There would be no Dymaxion house for the masses. As noted by Jay Baldwin, one of Fuller's most knowledgeable biographers, the house would have faced obstacles in the marketplace even if it had been commercialized.[39]

- "Construction trade unions made it clear that they did not intend to hook up Dymaxion houses that had been pre-plumbed and pre-wired" by aircraft workers.
- There was no "infrastructure of local dealers and trained installation crews."
- Conforming to building codes that made no allowance for such a radical house would have been very difficult.
- Banks would have been reluctant to lend "mortgage money to the buyers of houses that do not meet codes, however antiquated."

A man with extraordinary energy and an ego to match, Fuller was nothing if not resilient. Despite failures to launch two Dymaxion houses, the Dymaxion car, the modular bathroom, and other inventions, Fuller continued to invent and push boundaries of workable designs. The geodesic dome, his signature invention, was patented and built in endless varieties and at last gave Fuller commercial success and widespread acclaim.

Commencing the dome project after World War II, Fuller did much of the development work while teaching at Black Mountain College in Asheville, North Carolina, during the summers of 1948 and 1949. To bring his invention to market, Fuller founded Geodesics and licensed other companies to build domes according to his designs, some of which featured triangular and hexagonal components. The U.S. government and scores of companies built domes because—consistent with Fuller's design imperatives—domes promised to cover a given space with far less material and weight than any other shape and could be assembled in much less time than traditional structures. With their ultra-modern look, the domes also advertised that their owners were discerning consumers of a cutting-edge technology suitable for the jet age.

The first commercial dome was built in 1953 to cover the courtyard of Ford Motor Company's Rotunda building in Dearborn, Michigan. A few years later, industrialist Henry Kaiser built a large dome at his Honolulu Village resort in Hawaii. Spanning 145 feet and made of Kaiser aluminum, the dome was assembled in just 22 hours and housed a concert hall seating two thousand people. Taking advantage of the venue's unique acoustics such as pleasant reverberation, Arthur Lyman recorded many of his exotica albums there, including *Taboo*.[40] One of Geodesics' most ambitious domes was the American Pavilion at Expo 67 in Montréal; with a steel frame and two thousand acrylic pieces, it had a diameter of 250 feet (Figure 8.4). In addition to these iconic domes, several hundred thousand have reportedly been built throughout the world.

Owing mainly to the dome and Fuller's far-out ideas for other projects, he became a cultural and counter-cultural icon in the 1960s and 1970s, a guru from some future time and place. During these decades he traveled around the world giving lectures—some lasting more than five hours—and wrote about two dozen books, including the famous *Operating Manual for Spaceship Earth* (1968). As the world's foremost technological utopian, he told audiences that "man can create miracles."[41] On January 10, 1964, Fuller reached the pinnacle of popular success when he and his creations were featured on the cover of *Time* magazine, an image that appeared posthumously on a U.S. postage stamp in 2004 (Figure

Figure 8.4. Biosphère Montréal. Source: courtesy of Wikimedia Commons, Cédric Thévenet, photographer.

8.5). Fuller was also interviewed by *Playboy* magazine but (fortunately!) did not grace the cover.[42]

✳ ✳ ✳ ✳ ✳

Among accomplished inventors and entrepreneurs, Fuller's career was somewhat unusual. Many inventor-entrepreneurs begin with relatively simple inventions and eventually gain the experience and confidence to tackle more challenging projects. Arguably, Fuller's career trended in the opposite direction. His house and car projects were immensely difficult and might have initiated new technological traditions had they been commercialized.

The first Dymaxion house was a devilishly complex structure not nearly ready for production—likewise the Dymaxion car, even Model 3. These failures did not weigh heavily on Fuller, who rationalized that they enabled him to experiment with innovative design ideas. Apparently eschewing financial gain, he was not driven (as

Figure 8.5. Fuller and his inventions on a 2004 U.S. postage stamp.
Source: author's photograph.

are most inventors) to commercialize and profit from his novel technologies. And no investors, entrepreneurs, or companies stepped up with support for further development and testing. These radical designs were a cautionary sign to potential investors, especially during the Great Depression.

The later, and equally complex, Dymaxion house generated much public interest and even orders. In the postwar period, new materials for building the complex structure were available; the design was still radical but feasible, as Barwise showed. However, Fuller wasn't ready to finalize his plans, much less allow the project to reach the next stage. The unresolved disagreement between Fuller and his backers led to the project's demise.

Apart from the short-lived Dymaxion deployment unit, Fuller's only resounding commercial success was the geodesic dome, an eminently simple structure. Many articles and books celebrate Fuller's genius but downplay the early flops. The latter are in many ways more interesting than the dome, for they were highly innovative— totally unlike the technologies they were intended to replace. And in their very complexity they reveal Fuller's creative genius. An astonishing variety of domes built around the world also testify to the power of Fuller's pithy design imperative: do more with less. Unlike the Dymaxion houses and car, only the simple dome met this requirement and succeeded.

I wonder if Fuller was really indifferent to whether his creations were commercialized. If we take him at his word, he was more interested in inventing and designing technologies than in putting them into production. Yet throughout his entire career Fuller managed to swallow his scruples, seek backers, and launch companies— one of which commercialized the geodesic dome and furnished him with a substantial income.

✳ ✳ ✳ ✳ ✳

Of the three Dymaxion car prototypes, only Model 2 survives. It has been restored and resides in the National Automobile Museum in Reno, Nevada. In recent years two replicas have been built: one (after Model 2) by Norman Foster's team in England, the other (after Model 1) having been commissioned by the Lane Motor Museum in Nashville, Tennessee.[43]

Several Dymaxion deployment units can be seen at Camp Evans in Wall Township, New Jersey.[44] And perhaps there are others around the world.

The Dymaxion house prototypes have had an interesting history.[45] William L. Graham, a stockholder in the Dymaxion Dwelling Machine Company, bought the components of the two prototypes. In 1948, combining parts from both, he assembled his own Dymax-

Figure 8.6. Wichita house in the Henry Ford Museum. Source: courtesy of Wikimedia Commons, Michael Barera, photographer.

ion house in the country near Andover, Kansas, though it was never the family's principal residence.[46] Known as the Wichita house, the reconstruction was said to closely follow the original, though Fuller himself was disappointed. Many visitors remarked that the house resembled a flying saucer, whose form had been ordained in popular culture by science fiction novels and movies.

The Wichita house remained at that site for about four decades, but after standing vacant for some time became a raccoon habitat. Adding to the mess, the roof leaked. In view of the house's historical significance, the family had it disassembled and shipped to the Henry Ford Museum and Greenfield Village in Dearborn, Michigan. Museum officials were pleased to acquire the Wichita house, even though the Grahams had made alterations and many of the components—especially the aluminum—had deteriorated badly.[47] The house parts were passed along to the museum's restoration specialists for conservation. When this daunting job was completed, the house was reconstructed to approximate the Barwise prototype. In 2001 the Dymaxion house was put on display inside the museum (Figure 8.6), where it mutely testifies to Fuller's ingenuity and the technological optimism he personified.

NOTES

1 Baldwin (1996) and Pawley (1990) are generously illustrated overviews of Fuller's work.

2 Fuller quoted in Krebs (1983).

3 Fuller (1983:xvi).

4 Fuller (1983:vii–xxxii); Kelly (1982) also discusses Fuller's philosophy.

5 Tisdale (1930:44).

6 https://en.wikipedia.org/wiki/Sears_Catalog_Home, accessed 4 November 2016.

7 The aluminum was in the form of duralumin, a copper alloy.

8 My discussion of the house draws on Baldwin (1996:12–30), Fuller (1932, 1983:11–29), Marks and Fuller (1973:18–23), and Zala (1934). I rely mainly on Tisdale (1930), which reports on a Fuller show-and-tell uninflected by his later concerns.

9 Marks and Fuller (1973:21). Some sources give "action" as the origin of ion.

10 Fuller was inconsistent on some details and they vary among published accounts.

11 Tisdale (1930:44).

12 Fuller (1932:64).

13 Tisdale (1930:216).

14 Tisdale 1930:217).

15 Tisdale (1930:216).

16 MacLeish quoted in Marks and Fuller (1973:21).

17 Zala (1934).

18 Marks and Fuller (1973:21).

19 Fuller (1983:31); Hatch (1974:122–123); Marks and Fuller (1973:25). Pawley (2002) is a well-illustrated chapter on the Dymaxion car.

[20] Bush (1974:309).

[21] Fuller sought a patent for the Dymaxion car in October 1933, and a U.S. patent (#2101057A) was issued in December 1937.

[22] Marks and Fuller (1973:27).

[23] "Bullet-Shaped Auto Speeds on 3 Wheels," *New York Times*, 22 July 1933, p. 13.

[24] For a video of Fuller driving Models 1 and 2, see https://www.youtube. com/watch?v=YlLZE23EJKs, accessed 2 November 2016.

[25] "Dymaxion Automobile Unlike Anything Ever Produced in Motor Vehicle Line," *Los Angeles Times*, 30 July 1933, p. E4.

[26] Marks and Fuller (1973:112).

[27] On the disposition of Models 2 and 3, see Baldwin (1996:94, 99).

[28] https://en.wikipedia.org/wiki/Dymaxion_car, accessed 13 November 2016.

[29] Marks and Fuller (1973:32–34).

[30] Fuller, quoted in Hatch (1974:133).

[31] Marks and Fuller (1973:34–35).

[32] For U.S. wartime aircraft production, see https://en.wikipedia.org/ wiki/World_War_II_aircraft_production, accessed 8 November 2016.

[33] Marks and Fuller (1973:35).

[34] Information on the later Dymaxion house comes from Baldwin (1996), Hatch (1974:173–181), Marks and Fuller (1973:35–37), and Rybczynski (1992); http://hpef.us/publications/preserving-the-recent-past-publications/preserving-a-prototype, accessed 10 November 2016; http://www.docomomo-us.org/register/ fiche/ dymaxion_wichita_ house, accessed 10 November 2016.

[35] Marks and Fuller (1973:128–141) include dozens of photographs of component manufacture and assembly of the Barwise house. For a promotional video showing the Barwise house, see https:// www.youtube. com/watch?v=Vx5VJ1yd3HQ, accessed 17 November 2016.

[36] Both quotes from "Fuller's House: It Has a Better Than Even Chance of Upsetting the Building Industry." *Fortune*, April 1946, pp. 166–179.

[37] See Baldwin (1996:56–57) on the house's obvious flaws.

[38] On the orders see Baldwin (1996:52).

[39] Fuller's 1954 patent for the geodesic dome: #US2682235A. A geodesic dome must also include some pentagons.

40 https://en.wikipedia.org/wiki/Arthur_Lyman, accessed 13 November 2016; Marks and Fuller (1973:224–225).

41 Fuller quoted in Krebs (1983:1).

42 *Time*, 10 January 1964; *Playboy*, February 1972, pp. 59–71.

43 https://en.wikipedia.org/wiki/Dymaxion_car, accessed 13 November 2016; Glancey (2010).

44 "War Shelters, Short-Lived Yet Living On," *New York Times*, 2 January 2014, p. D1.

45 http://www.docomomo-us.org/register/fiche/dymaxion_wichita_house, accessed 13 November 2016, P. D1.

46 For a video showing the construction of the Wichita house in 1948 (the 1946 date is wrong), see https://www.youtube.com/watch?v=bh5IIkH4xkA, accessed 17 November 2016.

47 http://hpef.us/publications/preserving-the-recent-past-publications/preserving-a-prototype, accessed 13 November 2016.

9

THE NUCLEAR–POWERED BOMBER

World War II ended with Japan and Germany defeated—their people desperate—and no longer a threat to the United States and its allies. However, as an expansionist Soviet Union colonized Eastern Europe, Americans perceived a new and sinister threat. To keep the Soviet Union in check, the United States developed technologies of deterrence. These were in effect political technologies for addressing problems in international relations.

Members of the military–industrial–academic complex that flourished during the war offered a host of ideas for deterrence technologies such as nuclear submarines, ballistic missiles, and long-range heavy bombers. The most grandiose visions included nuclear-powered rockets and airplanes. The federal government initiated development programs, feeding lucrative contracts to corporations hungry for postwar work, but not all projects were successful. Among the most costly and embarrassing failures was the air force's nuclear-powered bomber.[1]

✳ ✳ ✳ ✳ ✳

The possibility of a nuclear plane was conceived during the war by Enrico Fermi, who collaborated with other physicists and technicians at the University of Chicago to create the world's first controlled nuclear chain reaction. After the war the vision of nuclear-powered flight fostered a community of advocates, a constituency that argued the need for this technology and urged government support. A project was established in 1946, the joint venture of the U.S. Army Air Forces (which in 1947 became the U.S. Air Force) and the National Advisory Committee on Aeronautics. Called Nuclear Energy for Propulsion of Aircraft (NEPA), the project was directed by Major General Curtis LeMay, the military's outspoken promoter of air power who oversaw the saturation bombing of Japanese cities. The chief engineer was Andrew Kalitinsky, who worked for Pratt & Whitney Aircraft and served in naval intelligence during the war. Both LeMay and Kalitinsky were trained as engineers.

The NEPA contracted with many corporations that had participated in the war effort, including nine makers of aircraft engines. The prime contractor was Fairchild Engine & Airplane Company. Several physicists who worked on the army's Manhattan Project and designed the atomic bombs dropped on Japan were enlisted as consultants. The project was headquartered at Oak Ridge National Laboratory in Oak Ridge, Tennessee, supplier of fissionable uranium.

Simple physics suggested the possibility of a nuclear-powered plane.[2] In addition to generating an enormous amount of radiation, a nuclear reactor produces heat in abundance. To prevent a meltdown, a reactor's core must be cooled but the heat removed can be put to work. In a commercial nuclear power plant, the heat creates steam that drives turbines coupled to electrical generators. In a nuclear airplane, the heat would raise the temperature (and thus the pressure) of air high enough to power a turbojet engine.

Working from front to rear, a turbojet engine consists of three major parts: a compressor, combustion chamber, and turbine. A single shaft runs the length of the engine, connecting the compressor and turbine. Air entering the front of the engine is compressed and, in the combustion chamber, mixed with fuel and ignited. Now under great pressure, the hot air drives the turbine and compressor. Exhaust from the turbojet provides thrust.

In a nuclear airplane, the reactor replaces the turbojet's combustion chamber. The compressor pumps air through the reactor's core, partially cooling the core while heating the air to a very high temperature. The hot air is returned to the engine where it rotates the turbine blades and drives the compressor. Again, thrust is supplied by the engine's exhaust. But this design (the "direct cycle" system) poses a problem: air passing through the reactor becomes radioactive, contaminating the engine's exhaust. An alternative design (an "indirect cycle" system) could avoid this problem by using a closed system in which an intermediate fluid—such as a molten metal—transfers the reactor's heat to a heat exchanger, which in turn supplies the hot air going to the turbojet. An indirect cycle propulsion system would be heavier and vastly more complex to design and build but in theory would not spew radiation.

Radiation inside the plane was a more serious problem because it would quickly kill the crew. In describing the NEPA in 1947, the *New York Times* claimed that "those who understand atomic energy" are certain that "the first atomic-powered aircraft will be a pilotless plane [because shielding] the pilot from the deadly radiation would require a weight that would make the plane of little use."[3] Kalitinsky, however, was confident that a shield could be designed to contain the radiation without making the plane too heavy. Aware of the radiation problem, which would also affect a plane's critical components and systems (from tires to electronics), many scientists doubted whether a nuclear plane was feasible. An outspoken critic of the NEPA was J. Robert Oppenheimer, the nuclear physicist who had been scientific director of the Manhattan Project. He was quoted in the press as saying that "nuclear power for planes . . . is so much hogwash."[4]

Despite objections by Oppenheimer and other distinguished scientists, advocates pushed ahead, touting a nuclear plane's projected performance characteristics. As Kalitinsky claimed in a 1948 article, a nuclear plane would combine "extremely high speed and almost unlimited range . . . [which might] make possible a supersonic bomber capable of flying around the world without pausing to refuel."[5] In view of the perceived Soviet threat, these capabilities might have made strategic sense in 1948. After all, a nuclear bomber based in the United States could reach any target in Eastern Europe or Asia nonstop. Another threat, which advocates raised from time to time, was the vulnerability of the United States should the Soviet Union develop a nuclear bomber first.[6]

Although basic physics allowed for the possibility of a nuclear airplane, the engineering challenges were immense. In order to produce a flow of hot air sufficiently swift to drive the turbine, the reactor would have to work at extreme temperatures—perhaps at 2000°F (hot enough to melt copper and soften steel). This placed a priority on developing a heat-resistant reactor and shield materials. In addition, to fit inside even the largest bomber, a reactor and its shield would have to be very compact. The use of thick lead or concrete shielding, which could weigh 75 tons, was ruled out.

The nuclear reactor was an infant technology in the postwar period, but engineers had considerable latitude in envisioning a compact, high-temperature reactor. However, calculations alone could not determine which reactor and shield designs would work best in an airplane. As in the case of many complex projects that push into new realms of engineering, the nuclear plane would require new principles of engineering science—lots of them. Laboratory tests were needed, for example, to measure the response of specific reactor, shield, and aircraft materials to high temperatures and to bombardment by neutrons and gamma rays. In the late 1940s, it would have been impossible to predict what these tests would reveal or even which materials might have to be tested. It was a matter of sheer faith that ingenuity and millions of federal dollars would yield a path for producing the necessary principles. And of course only ground and flight testing of actual reactors and shields—coupled to turbojet engines—could determine the real-world applicability of the new engineering science. Design of the airframe and associated components also presented many difficulties, but it was believed that the details could be worked out at a later stage.[7]

Clearly, no one in the late 1940s could have accurately predicted the time, money, and human resources it would take to solve the engineering problems and build a nuclear-powered bomber. And of course the eager engineers could not have foreseen the many ways in which their project would become embroiled in international and domestic politics.

Because even the most advanced shielding would still leak radiation, the crew's safety became a vexing problem, perhaps a showstopper.[8] Would radiation exposure hinder the crew's ability to perform their duties? Was there a limit to the number of missions the crew could fly? What long-term health risks did the crew face?

To answer these questions, the NEPA formed a Medical Advisory Committee. Scientists at this time believed that there could be "safe" doses of radiation. We know today that even tiny doses of radiation cumulatively raise the incidence of certain cancers at a population level. Because so little was known about radiation effects, the committee considered experimenting on healthy people, exposing them to increasing amounts of total-body radiation. Arguing against this move, Shields Warren—chief of the Atomic Energy Commission's (AEC) Division of Biology and Medicine—noted, "It was not very long since we got through trying Germans for doing exactly the same thing."[9]

The committee waffled on whether to proceed with human experiments, first approving them in principle but then deferring specific plans until alternatives were explored. A convenient alternative was to employ findings from clinical medicine. For several years the committee enlisted physicians to furnish information on the effects of therapeutic exposures to patients, including total-body irradiation. Owing to small sample sizes and lack of data on long-term effects, these studies left the questions unanswered. In the end the NEPA and other nuclear projects exposed people—both civilians and military personnel—to radiation. These were human experiments in all but name but they were rebranded as "occupational hazards," such as troops working in the fallout zone of a nuclear blast or the crew of a plane flying a working nuclear reactor.

Before the NEPA could conduct occupational experiments, the program had to assess whether building a manned nuclear bomber was feasible. Could the engineering challenges be overcome? What scale of effort might be required? Accordingly, the Department of Defense (DOD) asked the AEC to review current knowledge and make recommendations. The AEC turned to MIT to do the review. MIT, which among universities was becoming one of America's top military contractors, assembled a group of experts. Their puzzling

report, issued in September 1948, allowed for contradictory interpretations. On the one hand, the report suggested that nuclear-
powered flight was achievable in principle. On the other hand, it
forecast that this achievement would require fifteen years of effort
and an investment of $1 billion. If the U.S. government deemed the
nuclear bomber a necessity, then it could be done.

Known as the Lexington report, MIT's top-secret document was
pondered by the DOD. The Joint Chiefs of Staff finally concluded,
in early 1951, that the military did need a nuclear bomber. The air
force and AEC, while closing down the studies at Oak Ridge National Laboratory, established a joint project: the Aircraft Nuclear
Propulsion program (ANP). Contracts were awarded to General
Electric (GE) to develop the propulsion system (nuclear reactor
and turbojet engine) and to Consolidated Vultee Aircraft Company (Convair) to modify a B-36 heavy bomber so that it could accommodate nuclear propulsion. Although the overarching goal
was to develop technology for early flight, the project's immediate
objective was to create a propulsion system.[10]

The ANP survived for a decade, sustained by a large constituency of interested parties—especially the air force and AEC
but also corporations and advocates in Congress. In its rivalry with
the army and navy, the air force focused on this project as a way to
exhibit competence in nuclear matters. The army had succeeded
with the Manhattan Project and the navy was building a nuclear
submarine, and so the air force would have its nuclear bomber.

Supposedly a secret project, the print media carried enthusiastic
reports throughout the ANP's life.[11] In one illustrated article of
1957, a GE official described in some detail how ANP reactors
were tested.[12] In 1958 Major General Donald J. Keirn—the air
force official in charge of the ANP—was interviewed by *U.S. News
& World Report*.[13] He mentioned the ANP's major development
activities, even acknowledging some difficulties.

In response to changing political priorities and external technological developments, the project shifted back and forth among
major goals. Would the ANP be mainly a research project, or
would it strive to achieve a militarily useful bomber as Keirn insisted? Would the plane fly at supersonic or subsonic speeds? What
kind of nuclear weapons could be carried? Could it launch mis-

siles? Despite changing goals and objectives, the imperative to create a working propulsion system drove the majority of the project's activities from year to year.

In 1959 more than fifteen thousand people were working for the ANP's prime contractors and subcontractors.[14] The project ended up costing slightly more than $1 billion—not including administrative and management expenses at the AEC and DOD. Almost half of this total went to GE for contracts amounting to $527 million in pursuit of a direct cycle system.[15]

Hedging its bets, the ANP also funded work on the indirect cycle system—around $280 million—much of it going to Pratt & Whitney Aircraft ($164 million) and Union Carbide Nuclear Company, which ran Oak Ridge National Laboratory ($68 million).[16] Their difficult project failed to make a working reactor and heat-transfer system but did create much new engineering science.[17]

Work by GE and its many subcontractors was more productive. The first GE reactor, known as P-1, was built as part of an "early flight program," whose name reflected the air force's optimistic goal. Although containing conventional materials—such as aluminum and stainless steel—whose properties were well understood, the P-1 also embodied innovations that would be used in later reactors. The P-1 was supposed to lead to ground tests but the program, which began in 1951, was ended in early 1953.

Throughout the 1950s detractors, including officials in the Eisenhower administration, raised objections to the project—objections that intensified as the project dragged on with no end in sight. Other ambitious military projects were also clamoring for funds and the Korean War was still under way, and so in 1953 the Eisenhower administration tried to discontinue the ANP. Bowing to the constituency, however, the administration compromised with Congress, allowing the project to survive and enjoy a robust funding boost in the years ahead.

In 1957 the Soviet Union gave the ANP a gift by launching Sputnik, the first earth-orbiting artificial satellite. This event touched off a flurry of inflated claims that the United States was far behind the Soviet Union in the missile race and energized the ANP's constituency to pressure the Eisenhower administration to continue the project. Adding to the urgency, erroneous reports circulated that the Soviet Union had developed a nuclear bomber, but its pro-gram had in fact achieved little. By contrast, the American program was plodding along—despite periodic threats to end funding—in the context of a Cold War with the Soviet Union that had become a nuclear arms race.

✳ ✳ ✳ ✳ ✳

Other American weapons programs, in the meantime, were reach-ing their goals far more quickly than the ANP. Beginning in 1954 the Boeing Company flight tested a new long-range heavy bomber able to deliver nuclear weapons. The B-52 Stratofortress, which went into service in 1955, was—and still is—a formidable weapon. It can fly 8,800 miles without refueling and carry a payload of 70,000 pounds. With its eight Pratt & Whitney engines, the B-52 had a maximum takeoff weight of 488,000 pounds.[18]

Also in 1955 the *Nautilus,* the world's first nuclear-powered sub-marine, began operation.[19] Developed under the determined leader-ship of Admiral Hyman G. Rickover and built by General Dynamics, the *Nautilus* was the beginning of America's nuclear navy. There were fewer engineering constraints in designing a nuclear power plant for a submarine or other large ship than for an airplane. Sig-nificantly, because weight was not as critical, the Westinghouse re-actor was thoroughly shielded to protect the crew and the ship's components and structure. Although the *Nautilus* could not launch missiles, much less missiles carrying nuclear warheads, by the end of the decade other submarines had these capabilities.[20] Nuclear-powered submarines cruised beneath the sea for months at a time, furnishing another technology of deterrence after 1960.

During World War II Germany created the V-2, the world's most advanced rocket. After the war Wernher von Braun and dozens of other German "rocket men" came to the United States along with boatloads of V-2 parts. These imports of German human and mate-rial resources were the foundation of America's missile programs developed by the army, air force, and navy. By 1960 intercontinental ballistic missiles able to carry nuclear warheads were in the Ameri-can arsenal, joining nuclear submarines and B-52s to complete the deterrence triad.

While these weapons systems were being put into operation, ANP contractors conducted a plethora of experiments that created new engineering science in abundance. The effects of high temperature and radiation on materials that might be used in the reactor, turbojet engine, and airframe were studied and the results of experiments summarized in thousands of classified reports.[21] Among the important findings, it was learned that certain non-clay ceramic materials did not melt at high temperatures and could be used in a reactor.

GE's engineers also achieved many technical milestones in a series of heat transfer reactor experiments (HTRE) that coupled experimental reactors to one or two turbojets. The ground tests were carried out at the AEC's National Reactor Testing Station, established in 1949 to assemble and test reactors developed by federal programs.[22] The facility was located in a sparsely populated area of eastern Idaho. This isolation turned out to be a good idea because the experiments "shot shocking amounts of contamination into the Idaho sky."[23] In these experiments numbered (HTRE 1–3) reactors, controls, and turbojets were installed in a movable test stand that was towed on rails by a heavily shielded vehicle to and from the maintenance and assembly building to the test pad. The reactor sat on the test stand's upper level, connected by formidable ductwork to the turbojet on the lower level.

In January 1956 the first reactor, which used several novel materials, ran a modified GE J47 turbojet engine entirely on nuclear power. This was a noteworthy achievement that confirmed theoretical calculations (Figure 9.1). During 1956 HTRE-1 ran at high power for a total of 150.8 hours. GE claimed that HTRE-1, "if applied to a prototype aircraft propulsion system, would have made possible the flight of a load-carrying aircraft a distance of approximately 50,000 miles at intermediate subsonic speeds without refueling or touching down."[24]

HTRE-2 built on HTRE-1 using a modified reactor that contained many heat-tolerant materials, including ceramic fuel cartridges. HTRE-2 was operated "with fuel temperatures between 1200° and 1750°F and reactor exit-air temperatures between 875° and 1125°F."[25] In some tests fuel temperatures reached 2750°F. Be-

Figure 9.1. HTRE-1 and crew. Source: courtesy of the
National Reactor Testing Station.

ginning in July 1957, HTRE-2 ran at high power for more than one
thousand hours.

HTRE-3 had a full-scale reactor incorporating many ceramic
components. In addition, "the reactor was mounted horizontally
and was equipped with a flight-type shield."[26] Coupled to the reactor, two modified J47 turbojets reached full power in 1959 and operated for 126 hours. The first two HTREs had been started with
chemical fuel, but HTRE-3 achieved its first nuclear-only start in
December 1960.[27]

Having built complete propulsion systems, including controls,
the HTRE series showed conclusively that a nuclear reactor could
power one or two turbojet engines, perhaps more. But they did not
show that these huge and heavy propulsion systems could work in
an airplane, much less fit inside one. And then there was that pesky
radiation problem.

The J47 engine, which dated to the late 1940s and was the first American commercial jet engine, was not ideally suited for nuclear operation. The modifications helped but in 1955 GE began work on a larger reactor-compatible turbojet, called the X211. Instead of advancing to a proposed HTRE-4 with the new engine and latest reactor design, GE—responding to constantly changing directions from the air force and AEC—designed a radically reconfigured and streamlined propulsion system that in principle could actually fit inside a plane.

The HTRE systems had been ungainly monstrosities, but now GE engineers designed an integrated propulsion system—the XNJ140E—of considerable elegance. With the reactor and shield built into the turbojet, the air flowed from the compressor, through the X211 reactor, to the turbine and exhaust. One shaft connected the compressor to the turbine and ran through the middle of the re-actor's core (Figure 9.2). A design was also prepared in which one reactor served two turbojets.

Development of the XNJ140E system was approved as an objec-tive of the ANP, with December 1962 set as the target date for ground testing a prototype. An enthusiastic air force projected that

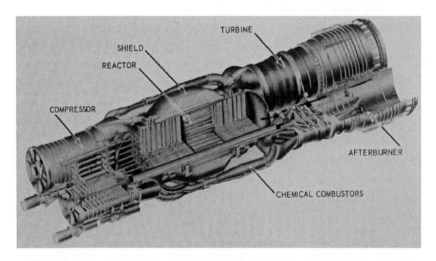

Figure 9.2. Cutaway drawing of General Electric's XNJ140E nuclear propulsion system. Source: General Electric (1962: Figure 4.1).

the system could achieve first flight in 1965. The new system was completely designed and critical components tested.[28] The ANP was apparently on the verge of success.

The plan was to install one or more XNJ140E systems in a large Convair plane, the NX2, designed for nuclear propulsion. This company was not new to flying reactors. From 1955 to 1957, Convair had flown a tiny nuclear reactor in a modified B-36H bomber, but it did not (and could not) power the plane. In this human experiment, the plane had two shields to protect the crew from radiation. The separable crew compartment had its own shield consisting of a layer of lead, 0.25 to 2.50 inches thick, and about a foot of rubber. Only the pilots could see out the small window, which was a sandwich of leaded glass and plexiglass almost a foot thick. Although giving some protection to the five-person crew, the split shield allowed for greater irradiation of airplane components—accelerating the deterioration, for example, of rubber and oils. And the plane's electronic equipment had to be placed in the shielded area of the crew compartment.[29] The split-shield system was based on the assumption that the two shields together would weigh less than a single effective reactor shield.

Taking off from Carswell Air Force Base in Fort Worth, Texas, the plane flew 47 missions over desolate areas in West Texas and New Mexico accompanied by a chase plane. In case of a crash, men would parachute out of the chase plane to render aid and secure the area, but there were no crashes. When the reactor was operating, sensors throughout the plane furnished data on radiation exposures of components and crew. After each flight the crew compartment was separated from the fuselage. The 35,000-pound reactor (not including shielding), housed in the plane's rear bomb bay, was removed and transported to a secure "hot" room. Technicians worked behind a six-foot-thick glass shield and employed remotely controlled robotic arms to handle the reactor's components.

In contrast to the Convair flight tests, the XNJ140E system could not be installed in a repurposed old bomber; the NX2 airframe had to be designed from scratch. The power plants would consist of several XNJ140E systems located in the rear half of the plane (Figure 9.3). Each swept-back wing would suspend a conven-

Figure 9.3. Artist's conception of Convair's NX2 nuclear bomber.
Source: General Electric (1962: Figure 1.2).

tional turbojet engine, which could be used if nuclear propulsion failed or if added power was needed for takeoff or fast cruising.

By 1960 both the Soviet Union and the United States had achieved the ability to destroy each other's civilizations and turn all of Earth's cities into smoldering rubble. These capabilities gave rise to the policy of "mutual assured destruction" or MAD, the belief that no rational person would authorize the first strike knowing that it would foment an all-out nuclear war that no side could win. It was in this context that the Eisenhower administration reappraised the relevance of the nuclear airplane.

With nuclear submarines, B-52s, and intercontinental ballistic missiles on line along with many American bases in Europe and East Asia, the original justifications for a nuclear bomber were no longer relevant. And the threat of a Russian nuclear plane had been debunked. So a new justification was invented: perhaps a nuclear plane could be used for spying.[30] But this justification was lame given that no nuclear plane would be ready to fly before 1965. By the late fifties the United States had spy planes (the supersonic SR-71 and the U-2) and spy satellites would perform the same function more effectively and less conspicuously after 1960.

In 1957 Eisenhower established the President's Science Advisory Committee, whose twenty or so distinguished members were mainly physicists and chemists. Many were chastened veterans of the Manhattan Project and were reluctant to urge development of new nuclear weapon systems, especially if they might accelerate the arms race. The committee did not view the ANP favorably, believing—in the words of its chairman George B. Kistiakowsky—that it was a "technical failure."[31] Accordingly, the committee recommended that the ANP be confined to basic research. This advice could have led Eisenhower to terminate the ANP but he didn't. Most likely he wanted the next president to take the heat from the plane's powerful constituency, which included not only the air force, AEC, and big corporations but also representatives from at least ten states where the ANP had facilities.[32]

Recognizing in 1959 that the ANP's survival was in danger, the air force took the unusual step of publishing in its own journal a series of upbeat articles detailing the nature and status of U.S. nuclear flight projects and focusing on the ANP and its importance for maintaining air superiority. Orchestrated by Keirn and written mainly by air force officials and engineers at major contractors, the articles in this special issue of *Air University Quarterly Review* claimed that significant progress had been made in tackling the ANP's engineering problems, but that more research and development were still needed—especially on the indirect-cycle system. The following year these same articles were turned into a book.[33] Perhaps this informative propaganda would influence policymakers to extend the life of the ANP.

In 1960 John F. Kennedy was elected president. Among his top advisers was Secretary of Defense Robert S. McNamara, a highly respected management expert. In search of expendable government programs, Kennedy charged McNamara with assessing the cost-effectiveness of various DOD efforts.[34] McNamara learned that to keep the ANP project going would require at least five more years of work and another billion dollars. In view of the project's uncertain outcome and the nuclear plane's lack of a realistic political mission, it was easy for McNamara to recommend the ANP's termination, for national defense would not suffer if it were discontinued. Previous attempts to end the ANP had ended in failure, but no longer would special pleading by powerful members of its constituency keep the nuclear bomber alive.

Following McNamara's recommendation, Kennedy sent a message to Congress on March 28, 1961, in which he stated that—despite fifteen years of effort and the expenditure of $1 billion—no nuclear aircraft could be built in the foreseeable future. Although as a distinct program the ANP was dead, Kennedy did authorize continued research on high-performance reactors and high-temperature materials.[35]

＊ ＊ ＊ ＊ ＊

Many scholars have tried to understand why this protracted and well-funded project failed to produce a working nuclear-powered aircraft. One line of thinking is that if the ANP had a single strong leader and a fixed goal, as did the navy's nuclear submarine program, the project would have moved along more swiftly and smoothly. A 1963 postmortem on the ANP by the General Accounting Office (GAO) gives some support to this idea.[36] The ANP was run by two large federal bureaucracies, the AEC and the air force. The lack of centralized management and one stable, long-term goal led—according to the GAO—to wasted money, unnecessary facilities, and other inefficiencies, but it did not prevent the growth of pertinent engineering science. Even the GAO's critical report touted the ANP's considerable accomplishments in this realm.

A few people opined that the ANP was doomed from the start because no amount of money and creative engineering could make it work. Researcher Ulrich Albrecht recently staked out this extreme position. "The infeasibility of nuclear propulsion for manned planes can be calculated on the back of an envelope."[37] His argument was that a reactor powerful enough to send aloft a big bomber could never be adequately shielded. Admittedly, building a nuclear airplane was one of the most challenging engineering projects attempted in the 1950s, but many engineers believed then (and still believe now) that it can be done. One argument in their favor is that today's largest planes, such as the Ukrainian freighter Antonov An-225, can carry a payload weighing more than a half-million pounds, perhaps sufficient to accommodate a reactor and adequate shielding.[38] A heavier plane would of course require strengthened landing gear and a longer runway because a nuclear plane, unlike a conventional jet, isn't constantly burning fuel and becoming lighter. (And of course today a nuclear bomber could be unmanned and operated remotely, eliminating the difficulty of shielding the crew.)

Even if the NX2 had been ready for flight testing in 1965, concerns about radiation polluting the environment—either in normal operation or in a crash—would have loomed large by this time and may have already doomed the ANP. Recall that GE's direct cycle system was a constant source of radioactive pollution. In the late 1950s, discussions were under way among the United States, United Kingdom, and Soviet Union on a treaty to ban nuclear testing. When it was signed in 1963, the treaty precluded testing and use of the GE propulsion system.[39] Even without this treaty, America's growing environmental consciousness—following the 1962 publication of Rachel Carson's book *Silent Spring*—likely would have ensured that the NX2 did not fly. So difficult were the engineering problems that Pratt & Whitney's indirect cycle system, which in principle did not pollute during operation, was nowhere near ready for ground testing and lacked a clear path to success.

To prevent radiation from escaping in a crash, engineers suggested many technological fixes. For example, if the crew expected an imminent crash, they could shut down and immediately jettison the reactor, which would parachute to a gentle landing. "Absurd" is the word that best describes this scenario because a hot reactor on the ground, which obviously lacked a cooling system, would release a radioactive cloud as it melted down.[40] And it is doubtful that less fanciful fixes, such as a heavily armored reactor, could prevent a radiation release in a crash. An ANP hazards committee pondered this problem in 1957, recommending that nuclear planes fly over the ocean, beginning and ending their flights on an island or coastal base.[41]

The nuclear airplane has become a zombie technology. Although ostensibly killed in 1961, it staggers on aimlessly. The vision of nuclear flight is so captivating that since 1961 several U.S. government agencies, including the National Aeronautics and Space Administration, have funded studies on reactor and airplane designs.[42]

Lockheed and other corporations have also sponsored studies insisting that the nuclear plane's time has come.[43]

Confident that an indirect cycle system can now be developed with modern engineering science and technologies, new constituencies in several countries are inventing justifications for ramping up large-scale projects—especially for civilian aviation. In a world concerned about global warming, for example, nuclear airplanes might be desirable because they would emit no greenhouse gases. Some advocates also claim that such planes would be a cost-effective alternative to conventional jets, both military and civilian.[44] At the present time, however, none of the new justifications seems sufficiently compelling to garner the necessary political support in a Western democracy. Of course a new program might require fifteen years of effort and tens of billions of dollars—and it too might fail.

The nuclear bomber project has left behind many tangible traces. In addition to copious classified and declassified documents, HTRE-2 and HTRE-3 survive at the National Reactor Testing Station near Idaho Falls, Idaho. More than a half-century after having shown that a nuclear reactor could power one or even two jet engines, these forlorn test stands—shorn of radioactive components—are monuments to the air force's unrealized dream.

NOTES

[1] For overviews of this project, see Bowles (2006); Colon (2007); Cortright (1995); Gantz (1960); Hewlett and Holl (1989); Tierney (1982). Snyder (1996) is an annotated bibliography of the project. On the postwar military–industrial–academic complex, see Leslie (1993).

[2] See, for example, Bussard and DeLauer (1965).

[3] Quoted in Graham (1947).

[4] Oppenheimer quoted in Bird and Sherwin (2005:433).

[5] Kalitinsky (1948:1).

[6] Garthoff (2016) discusses the politics of the U.S. and Soviet programs.

[7] For a discussion see Johnson and Cleveland (1957).

[8] Seren (1958a, 1958b) discusses safety issues.

[9] Warren quoted in Advisory Committee (1995:375).

[10] Project details are found, for example, in General Accounting Office (1963) and Lambright (1967).

[11] "Extraordinary Atomic Plane: The Fight for an Ultimate Weapon," *Newsweek*, 4 June 1956, pp. 55–60; Kenton (1956); Wendt (1951).

[12] Crocker (1957).

[13] "The Truth About a U.S. Atomic Plane: Interview with Maj. Gen. Keirn, Chief of Nuclear-Plane Project," *U.S. News & World Report*, 12 December 1958, pp. 38–39.

[14] Keirn (1960:17).

[15] General Accounting Office (1963:111–112).

[16] General Accounting Office (1963:112).

[17] General Accounting Office (1963:170, 183–184); Klingensmith and Lingenfelter (1960).

[18] http://www.af.mil/AboutUs/FactSheets/Display/tabid/224/Article/104465/b-52-stratofortress.aspx, accessed 30 January 2016.

[19] https://en.wikipedia.org/wiki/USS_Nautilus_%28SSN-571%29, accessed 30 January 2016.

[20] https://en.wikipedia.org/wiki/Ballistic_missile_submarine, accessed 30 January 2016.

[21] General Electric's studies are catalogued in General Electric (1962). This report was declassified in 1985.

[22] Department of Energy 2005. Stacy (2000, Chapter 13) discusses the lab's involvement in the ANP.

[23] Tucker (2009:134).

[24] General Electric (1962:41).

[25] General Electric (1962:49).

[26] General Electric (1962:50).

[27] General Electric (1962:57).

[28] General Electric (1962:63–71).

[29] Lewis (1958) supplies details on the Convair flights.

[30] See, for example, Joseph (1961).

31   Kistiakowsky quoted in Wang (2008:113).

32   Keirn (1960:18) mapped the involved states.

33   Gantz (1960).

34   Lambright (1967:28).

35   Kennedy quoted in General Accounting Office (1963:176).

36   General Accounting Office (1963).

37   Albrecht (1989:127).

38   https://en.wikipedia.org/wiki/Antonov_An-225_Mriya, accessed 16 May 2016.

39   https://en.wikipedia.org/wiki/Partial_Nuclear_Test_Ban_Treaty, accessed 15 February 2016.

40   See Seren (1958a:19).

41   General Accounting Office (1963:148). Discussions of hazards are also found in Gantz (1960).

42   See, for example, Fishbach (1969); Rom and Masser (1971); Schuwer (2015); Strack (1971); Wells et al. (2014).

43   Miller (1970).

44   See, for example, Frenkel (2008); Turba (2011).

10

DOMESTICATING THE BOMB

"Geographical Engineering" and Project Chariot

Seated before a group of reporters on August 6, 1945, President Harry S. Truman announced that sixteen hours earlier "an American airplane dropped one bomb on Hiroshima and destroyed its usefulness to the enemy." The atomic bomb, continued Truman, "had more power than 20,000 tons of TNT."[1] Although the vast scope of Hiroshima's devastation was still unknown, newspapers around the globe carried the story of America's atomic bomb, created in secret at a cost of $2 billion. Three days later a second bomb was dropped on Nagasaki and Japan soon surrendered. Pundits opined that depending on how the world's nations responded to this terrifying new weapon, the atomic bomb could end war or end civilization.

Although Truman described the Manhattan Project that built the bombs as "the greatest achievement of organized science in history," many participating scientists had conflicted feelings. On the one hand, they were proud of having harnessed the atom in the effort to defeat Nazi Germany, but the bomb hadn't been ready in time. On the other hand, many scientists were also appalled that their bombs had been dropped on a Japan already devastated and ready to sue for peace. These drastic actions immediately claimed more than two hundred thousand innocent lives; the bombs also flaunted America's military might—a blunt message to the Soviet Union—and exacted revenge for Pearl Harbor.

✳ ✳ ✳ ✳ ✳

Scientists were not the only people ambivalent about the atomic bomb. NBC radio commentator César Saerchinger remarked that "war is no longer against those who make war, but against the helpless, inarticulate, powerless mass of civilians. War with atomic bombs is a slaughter of the innocent, and hard to justify in patriotic terms."[2] More optimistically, Major George Fielding Eliot wrote in the *New York Herald Tribune* that "we may well be standing on the threshold of a true golden age of peaceful development," provided that "man can rise superior to the things he has created" and keep those things from the hands of persons with evil purposes.[3]

To prevent "those things" from falling into the wrong hands, some Americans argued that nuclear arms and materials should be placed exclusively under international control, perhaps through the United Nations. Other Americans believed that the United States should maintain its nuclear monopoly because countries such as the Soviet Union could not be trusted. Another faction argued that since the Soviet Union would inevitably develop its own bomb, both countries should cooperate and share nuclear information. Among these policy options Truman chose the one most politically expedient—recommending that Congress consider creating a commission to manage U.S. nuclear matters.

Responding to Truman's suggestion, in 1946 Congress passed the Atomic Energy Act, which established the Atomic Energy Commission (AEC). For three decades the generously funded AEC was dominated by a military mission: to develop and manufacture nuclear weapons. The AEC took control of all Manhattan Project facilities, including Los Alamos National Laboratory in New Mexico; Hanford Reservation in Richland, Washington; and Oak Ridge National Laboratory in Tennessee. Around this nucleus would grow a formidable and politically potent nuclear establishment, including additional national laboratories. The AEC's contracts with universities and corporations reached into most states, ensuring that elected officials would support its projects.

News of the bomb and its reliance on nuclear fission was widely reported in newspapers and magazines and on radio. Captivated by the fact that (ounce per ounce) U-235 packed three million times more energy than coal, apparently sober people conjured up visions of fantastic atomic technologies. In this brief silly season, physicist Lawrence M. Langer predicted that people would live underground where plants "will be grown under artificial heat and light," cars would be "powered by small chunks of U-235," and airplanes would be propelled long distances by the emission of particles from U-235.[4] Writing in the *New York Herald Tribune*, John T. O'Neill forecast that uranium would be a thousand times cheaper than coal and "atomic energy unquestionably will be made extremely cheap—like 'free air' at the service stations." He added that "locomotives operated by atomic energy will be practical in a short time."[5] In a CBS radio broadcast, Robert D. Potter ruminated on rockets, which if powered by atomic energy "would easily lift themselves, passengers, and freight into interplanetary space," enabling "transportation between planets that has always been called visionary. Today that vision is on the threshold of reality."[6] About the only technology not proposed was an atomic-powered toothbrush.

Alongside such fanciful visions, other nuclear futurists suggested somewhat more realizable technologies. Fissionable materials in nuclear plants, for example, would boil water, supplying steam to drive turboelectric generators for providing power. Other possibilities included the use of radioisotopes in environmental studies, biological research, and medical diagnosis and treatment. Such applications might allow the United States to be viewed—at home and abroad—as the benevolent provider of constructive new technologies, not merely the purveyor of nuclear destruction.

Propounding the theme of "Atoms for Peace," President Eisenhower addressed the UN General Assembly on December 8, 1953. Experts, he said, "would be mobilized to apply atomic energy to the needs of agriculture, medicine and other peaceful activities. A special purpose would be to provide abundant electrical energy in the power-starved areas of the world." This nation would "devote its entire heart and mind to finding the way by which the miraculous inventiveness of man shall not be dedicated to his death, but consecrated to his life."[7] Accordingly, the AEC promoted the development of nuclear power and the use of nuclear materials. The first nuclear-powered ship, the navy's submarine the *Nautilus*, was launched in 1954; the first commercial nuclear power plant came on line at Shippingport, Pennsylvania, in 1957. And many other proposed applications came to fruition.[8]

Eisenhower failed to mention that in the year before his speech the United States had developed a new kind of nuclear weapon. Physicist Edward Teller, a member of the Manhattan Project team at Los Alamos, had been working on the design of this bomb, known colloquially as the "Super." Instead of producing energy by the fission of U-235, the Super relied on the fusion of hydrogen atoms into helium. Nuclear fusion generated an enormous burst of thermal energy—and radiation—whose power dwarfed that of fission bombs. The Super became known as the H-bomb and begat a large and diverse arsenal of thermonuclear weapons.

Anxious to have a greater say in setting the country's nuclear research agenda, Teller lobbied the government to build yet another national laboratory. The AEC granted Teller's wish, founding the University of California Radiation Laboratory in Livermore, California. Teller resigned from Los Alamos and moved to the Livermore lab where he joined distinguished Berkeley physicist Ernest O. Lawrence, inventor of the cyclotron. Not surprisingly, Livermore's major mission was to develop and test thermonuclear weapons.

During the mid-1950s, when both the United States and Soviet Union were annually exploding dozens of bombs, fouling the planet with radioactive fallout, advocates for limits on nuclear testing began to gain a hearing in both countries. The dangers of nuclear fallout had become well known after the notorious *Lucky Dragon* incident. A Japanese boat, the *Lucky Dragon* was trolling for tuna on March 1, 1954, when the United States detonated a fifteen-megaton hydrogen bomb on Bikini Atoll. White powder drifting 85 miles east of the blast site dusted the 23 unlucky fishermen, causing radiation sickness and a year's hospital stay—and one fatality.[9] Closer to home, the United States also tested bombs in Nevada. Some ranchers claimed that the fallout had drifted off the test site, harming cattle and causing human illness. The AEC lamely denied the ranchers' plausible claims that the bombs were responsible.[10] By this time it was also understood that radiation was a cause of mutations and cancers that could grow decades after exposure.[11]

That the international community might adopt a moratorium or outright ban on nuclear testing was a dire threat to Livermore and Los Alamos.[12] How could these laboratories continue to refine bomb designs if they couldn't test them? Recognizing the risks, Livermore scientists led by Teller brainstormed and concocted another mission, one ostensibly in keeping with Eisenhower's theme of "atoms for peace." Soon began a new silly season of long duration.

The idea for the new mission took shape in 1957 after Egypt's nationalization of the Suez Canal and the ensuing Middle East war. Nuclear bombs, the Teller group concluded, could replace conventional explosives for monumental excavation projects, such as digging a new canal through Israel from the Mediterranean Sea to the Gulf of Aqaba. An even more ambitious—and dramatic—project would be a sea-level canal across the Isthmus of Central America, replacing the Panama Canal whose locks slowed ship traffic. The idea of using atomic bombs for engineering projects had been broached in 1950 by Los Alamos scientist Frederick Reines, but he was not optimistic owing to the "radioactivity hazard."[13]

Energized by their ambitious vision and downplaying the radioactivity hazard, Teller and fellow Livermore scientists Harold Brown and Gerald Johnson submitted a formal proposal to the AEC. They would create a program to reshape the earth with nuclear bombs, a program that Teller termed "geographical engineering." Nuclear excavation, they believed, would not only be simpler and quicker than using conventional explosives but would also be cheaper, ushering in a veritable revolution in earth-moving technology. In concert with industry, bombs "could excavate harbors and canals, stimulate the production of gas and oil, provide storage facilities for water or fuel, help gain access to deeply buried ores, create heat that could be captured for power production, and generate new atomic elements and isotopes for general use."[14] This program would ensure continuous design work at Livermore—and continuous testing. And it would deliver redemption for the bomb.

The AEC approved Livermore's Project Plowshare on June 27, 1957. Indeed, AEC chairman Lewis Strauss admitted that Project Plowshare had public relations value in excess of its value as public works. Beyond showing the Soviet Union that the United States had superior technologies for peacetime uses of the atom, Plowshare advocates believed it would help assuage anxieties about all things nuclear.

Nuclear war was especially on the minds of Americans during the 1950s when children practiced "drop drills" in school, cowering under their desks; some people dug backyard fallout shelters and stocked them with months of provisions; and radio companies placed two small civil defense triangles on tuning dials, which indicated stations that would remain on the air after a nuclear attack. In these fraught times, Plowshare—its backers believed—could lead to a more balanced assessment of nuclear risks and benefits.[15] Had Americans known the details of specific projects, however, it is more likely that they would have been aghast. Consider the Central American canal. "[O]ne alignment through Colombia would have required 262 nuclear bombs with an aggregate yield of 270.9 megatons."[16] The amount of radioactive fallout generated by so many explosions of such power would not have been trivial.

The AEC looked forward to Project Plowshare yielding a good return on a modest investment. In its first year, the project received only $100,000. Now director of Livermore, Teller pressed for more money and the AEC obliged. To reframe the bomb as a tool of peace, the AEC funded Plowshare at its peak at a level of $18 million annually.[17] These public relations expenses were pocket change: during the 1960s the AEC's annual budget was $2.4 billion.[18]

Although blasting a new canal across Central America remained Plowshare's glittering goal, the Livermore group grasped that a modest demonstration on American soil was needed first to furnish a proof of principle. A harbor was deemed the ideal project. Instead of soliciting proposals from maritime states or canvassing shipping companies, Livermore on its own sought a site that satisfied several criteria:

- Land owned by the U.S. government.
- No sizable human populations nearby.
- An economic justification for a harbor.
- Local stakeholders willing to assume post-excavation costs for constructing harbor facilities.

After studying maps and gathering cursory information on various possibilities, the Livermore group chose Ogotoruk Creek at Cape Thompson on the far northwest coast of Alaska, a traditional territory of Inupiat Eskimos. And so arose Project Chariot.[19] To sell Chariot to influential Alaskans, Teller himself went on a promotional tour—holding press conferences in Juneau, Fairbanks, and Anchorage, claiming that the harbor would stimulate the fishing industry and enable exports of coal. Lost in Teller's enthusiastic sales pitch were two unpleasant facts: Cape Thompson was icebound for all but three or four months of the year, and the coal deposits were hundreds of miles away and couldn't be exploited economically.[20]

Brushing aside Chariot's obvious negatives, Alaskan officials, newspapers, and several chambers of commerce—always eager to promote economic development—gave the project two thumbs up, perhaps unaware of Livermore's estimate that construction of the harbor facilities would cost between $50 and $100 million. Savvy businessmen suspected that local support for the harbor facilities would not materialize even if, as Teller himself insisted, construction costs could be kept to about $5 million.[21]

Many uncertainties surrounded Chariot's design. The basic idea was to bury bombs in the footprint of the proposed harbor, a channel one mile long that included a large turning basin. The explosions would vaporize vast amounts of soil and rock, sending the dust skyward and leaving behind craters that would fill with seawater. At this time, however, the engineering science of nuclear cratering was in its infancy. This is why the Livermore group could not accurately specify how many and how large the bombs needed to be and how deeply they had to be buried. More disquieting still, Livermore was unable to determine the total amount of radiation the blasts would release. Thus it was impossible to know how much fallout would be created and to what distances the winds might carry it. Despite these unknowns—seemingly fatal flaws—Livermore guesstimated that six thermonuclear bombs buried several hundred feet below the surface would do: four of one hundred kilotons and two of one megaton. Chariot's nuclear firepower would be 160 times greater than the Hiroshima bomb.[22] All the while Teller was touting progress in making a "clean" bomb, even pushing the idea to President Eisenhower, but a radiation-free bomb was the epitome of magical thinking.[23]

While Teller was touring, a Livermore contingent was checking out Ogotoruk Creek on the ground, preparing plans for the excavation. AEC contractors built an airstrip, brought in supplies by barge, and set up a temporary camp. And Livermore held additional public relations events in Alaska to bolster support for Chariot. Curiously, no one from Livermore bothered to inform the residents of nearby Inupiat Eskimo villages about the project.

After the economic justification for the harbor had been debunked, Teller and his colleagues revised the plan slightly and reduced the explosive power (Figure 10.1). Surprisingly, they also admitted that Project Chariot was in reality an experiment to obtain information about cratering that could inform, for example, the design of the new Central American canal. Decades later Gerald Johnson, who had been technical director of Chariot, confessed that "we just wanted to do an experiment. We didn't care if there was a harbor there or not."[24]

During a road show in Fairbanks, Teller addressed a group of University of Alaska faculty members, including biologists. Aware of the dangers of fallout, the professors queried Teller on the fallout expected from Chariot. Not willing to admit ignorance on this point, he simply asserted that "we have learned to use these powers with safety" and that the radiation release would be less than that of the cosmic ray background.[25] But the biologists knew that Teller's vacuous claim ignored potentially large radiation exposures at the local level, perhaps affecting flora, fauna, and the Point Hope Eskimo village—a mere 31 miles from ground zero. And the biologists were displeased by the absence of field studies on the local environment and the Eskimos' use of natural resources. Only after such information was acquired, argued the scientists, could an informed decision be made about whether to proceed with Chariot.

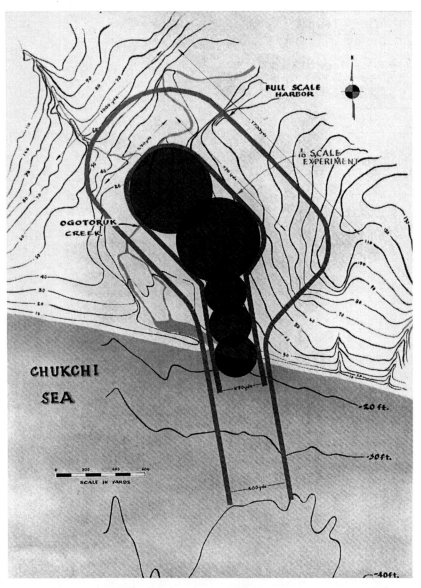

Figure 10.1. Draft plan for Project Chariot. Source: courtesy of Lawrence Livermore National Laboratory.

At the time collecting environmental information in advance of a land-modification project was a radical idea.

The yawning information gaps gave the University of Alaska an opportunity to tap into the AEC's bulging coffers. University President Ernest Patty, an ardent supporter of Chariot, requested that the AEC conduct pre-blast environmental studies. Faculty members in the biological sciences and wildlife departments also sent a letter to the AEC demanding thorough environmental research.

These pressures from the university and growing public awareness about the effects of fallout elicited a change in Livermore's strategy. No longer could Livermore claim that Chariot's fallout would be innocuous. After all, articles by luminaries such as Linus Pauling—Nobel laureate in chemistry—were documenting the hazards of fallout to living things.[26] To avoid suffering an utter loss of credibility, Livermore became a benefactor of environmental research at the University of Alaska, paying for extensive and unprecedented fieldwork on the region's geology, plants, and marine and terrestrial animals.[27] Significantly, geographer Don Foote was tasked with learning how the Point Hope people used the region's resources.

Although this suite of studies would delay Project Chariot, the Livermore group never doubted that the blast would occur eventually. Perhaps they believed that the generously funded environmental scientists could be counted on to support the project. According to Dan O'Neill, author of an important book on Project Chariot, a senior AEC official once "offered the Project Chariot Bioenvironmental Committee 'all the money you need for research' if the committee would just sign off on Project Chariot."[28]

Without so much as a single environmental study in hand, Livermore contracted in the summer of 1959 for the construction of a permanent camp at Ogotoruk Creek. Able to house seventy people, the camp "included ten dormitories, a mess hall, a shower and laundry building with hot water available at all times, and a building housing water purification equipment and three gasoline generators." Also built were warehouses, a communications building, and latrines. According to O'Neill, Project Chariot's infrastructure was in place and by August "the Chariot Camp was a bustling colony."[29]

While the "colony" (a word O'Neill had chosen carefully) was taking shape, the environmental scientists were conducting their first season of fieldwork. Don Foote obtained the cooperation of the Point Hope Village Council for his study of Inupiat cultural ecology. However, council members—who had learned about Project Chariot from Foote and others—resolutely opposed it. In a unanimous petition to the AEC, the council members stated that "we the undersigned the Point Hope Village Council do not want to see the explosion at the near area of our village Point Hope for any reason and at any time."[30] Although frustrated by the AEC's disregard of Point Hope residents, the Eskimos had more serious objections.

Contrary to the Livermore group's beliefs, the Ogotoruk region was not a vast white wasteland. In fact, its resources contributed greatly to the diet of the Point Hope people. Using the anthropological technique of participant observation, Foote lived with the Eskimos, ate their food, and accompanied them on hunting and gathering trips. He learned, for example, that the area was extensively used for taking caribou, collecting bird eggs, hunting seals, and so forth. The caribou hunt was especially critical to the Eskimos' survival. Foote's detailed records of Caribou kills showed that all animals were "shot within twenty-five miles of Ogotoruk Creek."[31] Had the Chariot blast occurred in March or April, the months preferred by Livermore, the Point Hope residents would have been unable to take caribou during a prime hunting time and famine would have followed.

In March 1960 three AEC officials, including a public relations specialist, at last toured Eskimo villages to campaign for Chariot. Curiously, none of the three was from Livermore. At the more distant villages, the AEC's "B-team" received no questions, which they naively interpreted as a sign of support for the project. The Point Hope people, who would be most directly affected by Chariot, were not so passive. Indeed, aware of the *Lucky Dragon* incident, they were deeply concerned about the effects of fallout on animals they relied on for food. After the presentation the village council again voted unanimously to oppose the blast, and the AEC again ignored the Eskimos' objections.[32]

A recording of the meeting made by the village council demonstrates that the AEC presentation, which included a film showing the simulated excavation of the harbor, was filled with falsehoods. The B-team claimed, for example, that

- Nuclear tests in the Pacific Proving Grounds had not rendered fish inedible.
- In just a few hours after the blast, Ogotoruk Creek would be largely free of dangerous fallout.
- Nuclear tests in the Nevada desert showed, through studies of cattle, that fallout would move harmlessly through the Arctic ecosystem.
- "Once Japanese survivors who received 'very great exposures' recovered from radiation sickness, they suffered no further effects."[33]

Dan O'Neill analyzed these and other claims in detail and found them to be either hopelessly misleading or untrue.[34]

Early in 1960 the AEC's in-house bioenvironmental committee issued a disturbing report. In specifying the conditions under which Chariot could be conducted, the report contradicted information it had already received from the University of Alaska scientists. In one egregious example, the report stated that "hunting activity is low" during spring.[35] The report also made erroneous claims about sea ice, snow cover, and patterns of bird migration. The university researchers furnished critical feedback and waited for the AEC committee to correct their report. But the AEC issued it to the press, unchanged. Outraged, environmental researchers Albert Johnson, William Pruitt, L. Gerard Swartz, and Leslie Viereck sent a strongly worded letter to the AEC drawing on the 1959 fieldwork to refute the report's claims. Jim Brooks, a researcher at the Alaska Department of Fish and Game, weighed in with a letter to the governor.[36]

Controversy over the report and the B-team's shameful performance at Point Hope marked a turning point, as opposition to Chariot spread from the affected Eskimos and disaffected environmental scientists to groups inside and outside Alaska. Indeed, opposition to Project Chariot helped catalyze growth of the modern environmental movement.

One of the first groups to express deep doubts was the Alaska Conservation Society, founded by Ginny Wood and Celia Hunter. After requesting an independent analysis of the raw environmental data, the society published in its *News Bulletin* a lengthy summary of environmental research from both the 1959 and 1960 field seasons.[37] This essay roundly refuted the AEC position that the blast would not imperil the Eskimos' livelihood.

Concerned about the complex relationship between science and politics in regard to nuclear affairs, Barry Commoner—a botanist at the University of Missouri—formed the Committee for Nuclear Information in 1958. The committee's newsletter *Nuclear Information* published two issues in the summer of 1960 on Plowshare and Chariot and invited the dissident Alaska scientists to submit articles. This they did in early 1961.

Pruitt's investigations, whose major findings he had earlier reported to the AEC, uncovered an ominous pattern in the Arctic food chain. Lichens, at the base of the chain, had an unusually high concentration of radioactivity from fallout. This radioactivity was in turn passed up the food chain to caribou and then to humans. Commoner took an uncommon interest in these findings, which energized his public advocacy for engagement with ecology. He also pointed out that AEC extrapolations of fallout from Nevada tests to Chariot were misguided because the ecosystems and food chains differed so dramatically.[38] Clearly, adding radiation to an ecosystem whose initial levels were already high would have been unwise.

About this time the AEC issued a summary report on Project Chariot, which downplayed Pruitt's findings and—extrapolating from Nevada Test Site and Pacific Proving Grounds data—claimed that the project's radiation effects "would be negligible, undetectable, or possibly nonexistent in areas distant from the excavation."[39] This report, along with articles in the Alaska Conservation Society's *News Bulletin* and Commoner's *Nuclear Information*, provoked discussions in national media such as *Science* and *Science News Letter*. These discussions drew in more groups strongly opposed to Chariot, including the Sierra Club and the Wilderness Society. Articles in their journals and in those of smaller societies called for an end to Chariot.[40] The AEC could do little to push back against these knowledgeable opponents, who had exposed the agency's flawed arguments and disingenuous claims. Drawing on the environmental and cultural studies, articles highly critical of Chariot began to appear in magazines.[41]

We do not know whether the opposition of environmental groups and adverse publicity were sufficient to end Chariot. What we do know is that the Inupiat Eskimos themselves were prepared to deliver a fatal blow with a time-honored American weapon: the assertion of property rights. La Verne Madigan, executive director of the Association on American Indian Affairs (a charitable organization), visited Point Hope and advised the village council on a new strategy for derailing Chariot. She pointed out that since the Eskimos had never ceded their lands to the U.S. government, the Organic Act of 1884 and Alaska Statehood Act of 1958 validated the Eskimos' claims to their traditional territory.[42] Thus the government lacked a legal basis for allowing the AEC to colonize Otogoruk Creek. Not surprisingly, the Eskimos threatened to sue.

With legal action imminent, environmentalists piling on, magazines publishing critical articles, and public sympathy growing for the residents of Point Hope, the AEC had become mired in a public relations debacle. Once set to launch Plowshare on a path to greatness, Project Chariot had become a distraction—even a liability. But Plowshare itself was still going strong and enjoyed the AEC's continuing support. What's more, Livermore was planning many new projects for "geographical engineering," rendering Chariot superfluous. The Livermore group moved on to greener pastures and Chariot was allowed to die quietly.

Dan O'Neill's use of the word "colony" to describe AEC's camp at Otogoruk Creek is especially apt. In the narrowest sense, a colony is a settlement that people impose on lands they do not own, usually against the will of indigenous inhabitants. In most cases colonizers regard the indigenous peoples as an inferior race or culture—a convenient rationale for extracting their resources, coercing them into agricultural or industrial labor, despoiling their lands, forcing them to relocate, and so forth. Many other places along the coast of Alaska might have been more economically suitable for blasting out a harbor, but Livermore chose to colonize an area inhabited by Native Alaskans.

The AEC was accustomed to riding roughshod over indigenous peoples. During the late 1940s and early 1950s, the AEC had displaced native Pacific Islanders to make way for tests of massive nuclear bombs, falsely claiming that the people could soon return home. Misled, the islanders offered little resistance. The AEC took it for granted that it could also appropriate traditional lands for its own purposes in Alaska, still a U.S. territory at the time planning began for Project Chariot. Unlike the Pacific Islanders, the Point Hope Eskimos and their local, regional, and national allies formed a constituency of environmental groups that resisted the AEC.

In the aftermath of Chariot's demise, the Alaskan scientists' radical idea began to take root as public policy: before a federal land modification project can be carried out, the agency has to acquire reliable information on the natural environment and on any potentially affected peoples. Such information would provide a sound baseline for assessing a project's likely impacts. This far-reaching ideal became codified less than a decade later in the National Environmental Policy Act of 1969.

It would be easy to dismiss Project Chariot as a harebrained scheme destined to fail that should have been stillborn at Livermore. Yet such a judgment ignores the fact that in sheer mechanical terms, nuclear earthmoving was—and is—technically possible. But turning possible into actual has a prerequisite: detailed planning must be done on the basis of pertinent engineering science. The members of the Livermore group were well aware that they had gone to Alaska with a less than half-baked plan, for they lacked the necessary scientific knowledge. Chariot was merely an experiment that would begin to provide such knowledge for designing future projects.

The failure to launch Chariot meant that other blasts would have to fill the knowledge gaps. And so, from December 1961 until May 1973, Plowshare detonated thirty nuclear bombs and ten conventional explosive devices at the Nevada Test Site. In addition, the project did nuclear tests at Carlsbad and Farmington, New Mexico, as well as Rifle and Grand Valley, Colorado.[43] On the basis of knowledge accumulating from these tests, Livermore designed dozens of new Plowshare projects at sites from California to West Virginia. Although preliminary work had been done at many sites, in some places provoking local opposition, no nuclear blasting ever took place.[44] Having failed in almost two decades to accomplish a single successful land modification project, Plowshare was put out of its misery in 1975.[45] There would be no new Central American canal.

※ ※ ※ ※ ※

In 1977 President Jimmy Carter, peanut farmer and retired nuclear engineer, folded the disreputable AEC into the new Department of Energy, which has a broad mandate to promote diverse energy sources. Livermore (now Lawrence Livermore National Laboratory) and Los Alamos National Laboratory continue to design new nuclear weapons, "testing" them in computer simulations.

When Project Chariot's permanent camp was abandoned, AEC contractors left behind a big mess: deep test holes, structural remains, construction equipment, pipes, wire, diesel fuel, radioactive tracers, and sundry scars on the landscape. Prodding by the Point Hope Village Council resulted in several episodes of remediation, including the removal of soil contaminated by diesel fuel and radioisotopes. Contractors scooped up many tons of soil, loaded them into large bags, and placed the bags in shipping containers for transport and final disposal (Figure 10.2, note the shipping containers).[46] Completed in 2014, the cleanups added still more scars, most of which will remain visible to future archaeologists.

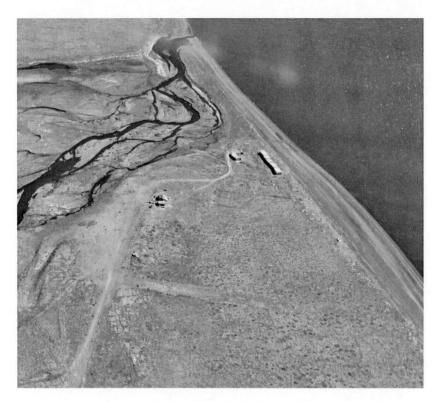

Figure 10.2. Aerial image showing damage to Ogotoruk Creek area. Source: courtesy of Wikimedia Commons, photographer Rtones, July 2014.

NOTES

[1] For quotes from Truman's speech, see http://www.pbs.org/wgbh/americanexperience/features/primary-resources/truman-hiroshima/, accessed 11 December 2016.

[2] Saerchinger quoted in Wendt and Geddes (1945:163).

[3] Eliot quoted in Wendt and Geddes (1945:166).

[4] Langer in PM, a short-lived New York newspaper, quoted in Wendt and Geddes (1945:178).

[5] O'Neill quoted in Wendt and Geddes (1945:186, 186–187).

[6] Potter quoted in Wendt and Geddes (1945:181).

[7] For quotes from Eisenhower's speech, see https://www.iaea.org/ about/ history/atoms-for-peace-speech, accessed 29 December 2016.

[8] Seaborg and Corliss (1971).

[9] Engel (1958).

[10] Berry (1957).

[11] On the debates over fallout hazards, see Kopp (1979).

[12] Kirsch (2005:12).

[13] Reines (1950:172).

[14] Kaufman (2013:2).

[15] The public relations blitz included articles appearing in *Popular Mechanics, Popular Science, Life, Time, Scientific American, This Week Magazine*, etc.

[16] O'Neill (2007:26).

[17] O'Neill (2007:28).

[18] Buck (1983:34).

[19] For my account of Project Chariot, I draw liberally on O'Neill (2007) and Kirsch (2005); see also Kaufman (2013).

[20] Kirsch (2005:48–49).

[21] O'Neill (2007:38).

[22] O'Neill (2007:45).

[23] Magraw (1988).

[24] Johnson quoted in O'Neill (2007:43).

[25] Teller quoted in O'Neill (2007:39).

[26] Pauling (1958); see also Lewis (1957).

[27] On the 1250-page final environmental report, see Wilimovsky and Wolfe (1966).

[28] Robert Rausch quoted in O'Neill (2007:162).

[29] Quotes in this paragraph are from O'Neill (2007:103).

[30] Excerpt from council letter quoted in O'Neill (2007:116).

[31] O'Neill (2007:182).

[32] O'Neill (2007:140).

[33] O'Neill (2007:141).

[34] O'Neill (2007:142–162).

[35] O'Neill (2007:175).

[36] O'Neill (2007:176-178.

37 O'Neill (2007:202–203).

38 O'Neill (2007:231).

39 U.S. Atomic Energy Commission (1960:55).

40 On people and organizations opposed to Chariot, see O'Neill (2007:233–247).

41 See, for example, Brooks and Foote (1962).

42 O'Neill (2007:250–251).

43 These data come from U.S. Department of Energy (n.d.).

44 About these obscure projects, see Beck, Edwards, and King (2011).

45 Kirsch (2005:7).

46 Tanik Construction and Fairbanks Environmental Services (2014).

11

CHRYSLER'S TURBOJET–POWERED AUTOMOBILE[1]

In the 1930s no person had ever heard the whine of a jet overhead because planes were powered by piston engines and propellers. The first hint of change came on August 27, 1939, when Germany's jet-powered He 178 made its inaugural flight. At that time teams in England, Italy, France, and the United States were also working on jet (turbine or turbojet) engines. Widespread interest in this new technology stemmed from theoretical calculations showing that the speed of piston-driven planes would soon plateau at an ultimate limit—around 400–500 miles per hour. By contrast, the turbojet engine was shown to be capable of crossing that barrier easily with no limit in sight.[2]

As World War II approached, governments grasped that faster fighter planes would have an immense advantage in battle and began to subsidize turbojet projects. Late in the war, Germany put a fighter jet into production; it had no effect on the outcome but pointed to the future of aerial combat. After the war development of military jets proceeded rapidly, especially in England, France, Soviet Union, and the United States. In 1949 North American Aviation began manufacturing the F-86 Sabre, which saw combat in the Korean War pitted against the Russian MiG-15s of North Korea.[3] And by the early 1950s experimental jet planes in several countries had broken the sound barrier (761 miles per hour = Mach 1). In the media and in popular culture, the jet came to epitomize the modern age.

The rapid transition of the turbojet engine from theory to capable power plant gave engineers an opportunity to envision new technologies. If a jet engine could revolutionize aircraft, then perhaps it could also revolutionize automobiles. This vision captivated many engineers because a turbojet engine's operating principles are simple. Air enters the engine where it is compressed and mixed with fuel. The air-fuel mix is ignited in the combustion chamber and expands greatly. Under enormous pressure the hot gases rotate the turbine (which drives the compressor) and pass through the exhaust nozzle, furnishing the thrust that pushes the plane. Unlike the piston engine, a turbojet does not have to convert reciprocating motion to rotary motion.

Although the turbojet was simple in principle, adapting it for the automobile entailed ceaseless design and material complexities. Such open-ended challenges require engineers to push hard on technology's frontiers. They may be confident in eventual success, but they also know that even a failed project may generate new engineering science and lead to publications and patents that will enhance their professional stature. Clearly, ambitious engineers—especially in the 20th century—have multiple incentives for promoting, and signing on to, the most challenging projects.

Engineers entranced by the vision of jet-powered cars had to convince executives in automobile corporations that the turbojet would be the engine of the future. Perhaps fearing that their companies might fall behind technologically, executives in many countries accepted the challenges and established turbine engine projects.[4] Rover was the pioneer with General Motors not far behind, but Chrysler also sustained a long-term project.

Drawing on its experience making some of England's earliest jet engines for aircraft, in 1950 Rover became the first company to adapt a turbojet for an automobile.[5] For two decades the company developed a succession of engines and installed them in test cars; a few even raced at Le Mans. Rover also created a turbojet for trucks but it did not go into production. Although markets were found for Rover engines, including auxiliary power generation, the automobile was not one of them. British Leyland subsequently acquired Rover's turbojet business and in 1972 sold it to Lucas Aerospace, which ended the automobile program.

General Motors, whose Allison Division worked on aircraft turbojets during the war, drew on this expertise to design automobile engines, placing them in futuristic cars.[6] Displayed at shows, the cars had handmade bodies with fins and bubble canopies that evoked the cartoonish offspring of a rocket and jet. The company's engines were designed for maximum power, not niceties such as fuel economy and engine life. Firebird I debuted in 1953 with a four-hundred-horsepower engine that could power the car up to two hundred miles per hour (Figure 11.1). Several new and equally bizarre Firebirds followed in later years, but they were also unsuitable for driving on city streets.

The General Motors team made splashy concept cars that attracted press coverage, as well as visitors to car shows who might glimpse the company's more conventional offerings. The company also built promising concept trucks and buses that employed their

Figure 11.1. General Motors Firebird I concept car, 1953.
Source: courtesy of Wikimedia Commons, photographer Karrmann.

powerful engines. But General Motors had no interest in creating turbojets that could be mass-produced for passenger cars. In the United States, Chrysler had this field to itself.

The Chrysler Corporation was founded in 1924 and had a reputation for competent engineering, but until the Airflow cars of 1934 it was not known for being conspicuously innovative.[7] The aerodynamic Airflow cars had radical styling that discouraged the average buyer and were "a failure so complete that it would cast a pall over Chrysler Corporation design for the next 20 years."[8] Perhaps the turbojet engine could lift this pall and help establish Chrysler as an innovator in postwar automobile technology.[9]

The major proponent of turbojet cars was George J. Huebner, Jr., a leading Chrysler engineer who acquired expertise in turbine design while developing a supercharged piston aircraft engine for the army air corps. In a supercharged engine, the exhaust-driven turbine compresses the incoming air, promoting more complete combustion and fostering greater power. Huebner gained further experience after the war when Chrysler built a lightweight turboprop engine for the navy. In contrast to a typical jet engine, the turboprop's compressor is geared to a propeller; the exhaust provides only limited thrust. These government projects gave Chrysler and Huebner a solid foundation in turbine design.

But would this experience be enough? Huebner recalled skeptics questioning whether a turbine engine could meet an automobile's demanding performance requirements, such as reasonable fuel economy. In an airplane a turbojet operates at peak fuel economy, which occurs at about 70 percent of maximum power. In an automobile a turbine engine would operate mostly at low power, especially in stop-and-go city driving, and guzzle fuel.[10] Then there was the matter of materials. Could a turbojet be built with readily available materials? What about engine braking? No passenger car could be built without it. Was it possible to build an engine small enough to fit into a car's engine compartment? Could a turbine engine, even if mass-produced, be competitive with a piston engine on a cost and performance basis? These were just a few of the issues that confronted Chrysler engineers in 1949 when they began to assess the feasibility of a turbine project. Huebner was confident that through "careful, thoughtful, and intelligent planning" these problems would be solved.[11]

A well-planned, creative engineering project promised to bring about an automobile revolution. The elegant, all-rotary-motion turbine engine would be vibration-free, quiet, and easy to maintain owing to its few moving parts. There would be no radiator, no distributor, and only a few bearings to lubricate. And it would weigh less than a piston engine of equal power and use almost any fuel. Enamored by the turbojet's potential, the charismatic Huebner convinced Chrysler executives to support the project.

From the project's beginnings in 1951, Chrysler engineers wanted to build an engine that could be mass-produced at reasonable cost for ordinary cars. Their turbojet would not be a specialty engine reserved for luxury, racing, or show cars. However, a decision to move forward on mass production could not be made until the turbojet engine lived up to its promise in the laboratory and on the road. More than this, Huebner and his team knew that no matter how well the turbine engine performed, Chrysler's executives would decide its fate.

Over the next several decades Chrysler built many turbine engines, each generation performing better than its predecessor.[12] Along the way, following Huebner's methodical approach, the Chrysler team conducted thousands of laboratory experiments and created a vast body of new engineering science relevant to turbine design. On the basis of this work, Huebner and his team members published articles and received many patents.

On March 23, 1954, James C. Zeder—Chrysler's vice president of engineering and research—publicly announced the birth of the first-generation turbojet engine. It was installed in a sporty 1954 Plymouth Belvedere and exhibited to the press. Zeder claimed that the turbine was more powerful than a comparable piston engine yet had about the same fuel economy (15–20 miles per gallon).[13] When Chrysler dedicated new proving grounds in Chelsea, Michigan, the company put on a show-and-tell with the turbojet Plymouth, racing it past a gaggle of reporters and photographers.[14] Images show a nattily attired Huebner "behind the wheel or standing next to the car—always grinning."[15] Justifying Huebner's obvious pride, the first turbine engine was a noteworthy accomplishment though not ready for consumers.

The engine spun at a maximum of about 45,000 revolutions per minute, far too rapid for powering accessories, transmission, and drive train. Accordingly, power for the generator (which also served as the starter motor) as well as the oil and fuel pumps was taken, through reduction gears, from the front of the compressor shaft.[16] The other end of the engine—the power turbine—was coupled, also through reduction gears, to the transmission.

Ordinarily, a turbojet produces very hot exhaust of around 1200°F, a danger to anyone and anything nearby. Significantly, such hot exhaust indicates an enormous amount of wasted energy, sapping power and reducing fuel economy. To ameliorate this problem Chrysler engineers devised a regenerator that fed exhaust gases,

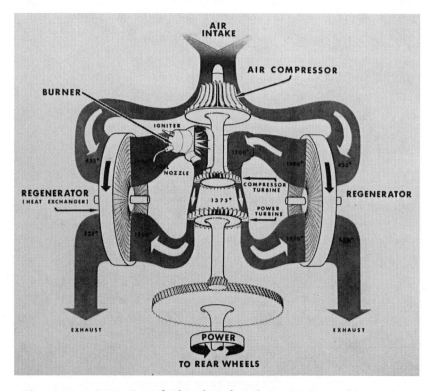

Figure 11.2. Operation of Chrysler's fourth-generation turbine engine. Source: Roy, Hagen, and Belleau (1964:3).

through heat exchangers, into the combustion chamber (Figure 11.2). This raised the inlet temperature to the power turbine and yielded much higher efficiency. The regenerator (sometimes called a "recuperator") greatly improved fuel economy and also reduced the exhaust temperature to less than 200°F at idle.[17]

Engineers had not yet solved the problem of materials. The engine used some strategic metals, such as cobalt, that were expensive and at times in short supply. A high priority in the design of later engines was to create heat-resistant alloys mainly using abundant metals.

Driving their turbine car around Detroit for two years, Chrysler engineers encountered few problems. Confident in the engine's durability, they were eager to test it on a long trip and publicize the company's accomplishments. Thus the engine, which was slightly modified from the 1954 version in part to increase power, was installed in a 1956 Plymouth.

Seeking maximum publicity, Chrysler engineers decided to drive from New York City to Los Angeles. Departing on March 26, 1956, Huebner did most of the driving, occasionally trading off with Zeder and other engineers. The car was accompanied by a sedan, truck, and three station wagons.[18] These vehicles carried technicians, tools, and spare parts that proved invaluable because on two occasions the engine had problems and only Chrysler people could repair it. Including down time for repairs, the trip took four days at about 40–45 miles per hour. Fuel consumption, probably diesel and kerosene, was 13–14 miles per gallon, far below Zeder's claim. (My 1956 Plymouth, with a 277 cubic inch V-8 and automatic transmission, routinely got sixteen miles per gallon on the highway.)

Chrysler's public relations people did their job well. Pulling up to Los Angeles City Hall at just before 9:00 A.M. on March 30, the caravan received a hero's welcome at a press conference attended by "more than a score of automobile officials, nearly a hundred cameramen, and Acting Mayor John T. Gibson."[19] The car's arrival was reported on the front page of the *Los Angeles Times*, which included a picture of Huebner peering into the car's engine compartment. In an interview, however, Zeder dampened expectations that the engine would be in production soon—perhaps in eight years, he said, but even that was admittedly a guess.

Writing years later, Huebner claimed that this trip "even then offered visible proof of the practicality of the gas turbine-powered passenger car."[20] But much work remained for the engineers before everyday drivers would regard the turbine car as practical. One nagging problem was accelerator lag. When the accelerator was depressed, it took about eight seconds before the engine revved up enough to move the car.

After their car's triumphal trip to California, Huebner and his team headed back to the laboratory to develop new generations of turbine engines having greater horsepower, better fuel economy, and briefer accelerator lag. To increase power it was necessary to raise the temperature (and thus pressure) of the gas driving the power turbine. This was accomplished with newly developed heat-tolerant alloys, whose exact compositions at the time were secret.[21] In the third-generation engine, completed in 1961, the power turbine's inlet temperature reached 1700°F.[22]

Chrysler engineers also devised a way to achieve engine braking. The accelerator pedal worked a linkage that changed the angle of the nozzles, which aimed the hot gases at the blades of the power turbine. The changing angle varied the speed and slowed the engine when the pedal was released (Figure 11.3).

In 1961 Chrysler believed it was time to carry out a new publicity blitz. An experimental sports car, called Turboflite, was equipped with the third-generation engine and "shown at auto shows in New York City, Chicago, London, Paris, Turin, and other cities."[23] Also paraded from show to show were a turbine-equipped 1960 Plymouth and 1.5-ton Dodge Truck. And of course another cross-country road trip was in order, this time in a 1962 Dodge. According to Huebner, inspection of the engine after this trip found it to be in "excellent condition."[24] What's more, the car's fuel economy on the trip supposedly matched that of a conventional car, perhaps fulfilling Zeder's earlier claim.

Eager to elicit consumer reactions, in spring 1962 Chrysler exhibited turbojet cars at Plymouth and Dodge dealerships in about ninety major cities across the United States and Canada.[25] Prone to

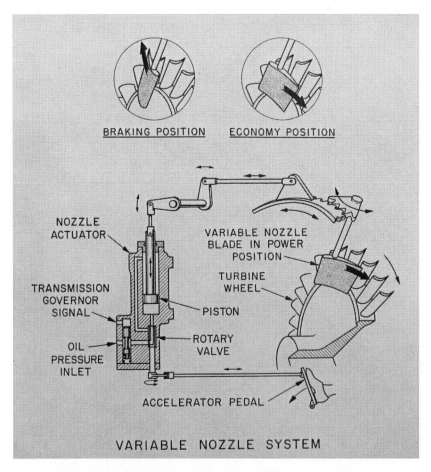

Figure 11.3. Chrysler turbine's variable nozzle system.
Source: Chrysler Corporation (1963:7).

exaggerate at times, Huebner reported that millions of people saw
the cars and more than thirteen thousand visitors were given rides.
When people were asked whether they would buy one, the response was heartening: 30 percent said they "would definitely buy
one" and another 54 percent said they would give it serious consideration.[26] Such polls are valuable public relations tools but have virtually no predictive power.

In the meantime Chrysler executives authorized an even more
dramatic display of confidence in their new engines, one that seemingly presaged true mass production. On February 14, 1962, the

company announced that it would build fifty turbine cars and loan them to members of the public. Other companies were shocked that Chrysler was making this bold move. Would Ford and General Motors, for example, have to catch up if the turbine car was successful?[27]

The new car had no special name other than the Chrysler Turbine. Designed by Chrysler, the Turbine's sleek bodies were handmade by Ghia, a Turin-based company that built Chrysler limousines and concept cars. A two-door hardtop with rear-wheel drive, the car was painted "turbine bronze" and had an interior to match. Inside and out, the Turbine was appointed with features such as taillights and wheel covers that alluded to turbine blades (Figure 11.4). It also had power steering, power brakes, and electric

Figure 11.4. Two views of the 1963 Chrysler Turbine.
Source: courtesy of the Transportation Division,
National Museum of American History, Smithsonian Institution.

SPECTACULAR FLOPS

windows—but no air conditioning. That the car vaguely resembled recent Ford models was no accident: Chrysler had lured designer Elwood Engel from Ford and the car was his handiwork.[28] The futuristic Turbine was distinctive but not a radical concept car in appearance.

Packed in huge wooden crates, the Ghia bodies began arriving at Chrysler in early 1963.[29] The first five cars gave Chrysler engineers an opportunity to perfect assembly-line techniques for installing the fourth-generation engine (Figure 11.5), modified TorqueFlite

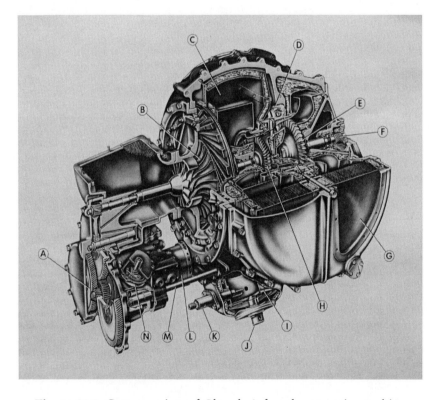

Figure 11.5. Cutaway view of Chrysler's fourth-generation turbine engine. Source: Chrysler Corporation (1963:2). Key: A = accessory drive; B = compressor; C = right regenerator rotor; D = variable nozzle unit; E = power turbine; F = reduction gear; G = left regenerator rotor; H = gas generator turbine; I = burner; J = fuel nozzle; K = igniter; L = starter-generator; M = regenerator drive shaft; N = ignition unit.

three-speed transmission, and attaching the body to the frame.[30] They also tweaked the styling, finally choosing black for the vinyl roof. None of the first five cars was part of the Fifty Car Program; the cars destined for consumers arrived later in the year.

In addition to its unique color and exterior styling, the Turbine had several unfamiliar dashboard gauges. One of them indicated the inlet temperature of the gases (up to 1800°F) as they entered the power turbine. The tachometer topped out at sixty thousand revolutions per minute. When running, the car sounded to outside observers like a distant jet engine, but the inside was quiet—especially at highway speeds.[31] Unfortunately, the Turbine was distressing to dogs because it emitted painful high-frequency sounds beyond the range of human hearing. Huebner guesstimated the cost of building the Turbine at between $50,000 and $55,000 each, almost twenty times the price of a 1962 Plymouth. But they weren't for sale.

The Chrysler Turbine debuted in New York City on May 14, 1963. Shown to the press at Essex House, reporters were later given an opportunity to go for a drive at Roosevelt Raceway. Chrysler dealers also saw the Turbine on display at the Waldorf–Astoria Hotel. It is easy to imagine the dealers' questions. When would the Turbine be available? How much would it cost? Would the engine be an option on other Chrysler cars? Would dealers need special equipment to repair the engines? At that time Chrysler could not have given firm answers to any such questions.

The Turbine's rollout was marred by a publicity stunt that didn't go as planned. With Huebner at the wheel, several Chrysler men attempted to drive the Turbine across Manhattan during rush hour. At one point the engine stalled, causing a massive traffic jam at Third Avenue and 40th Street. Trying in vain to restart the engine, Huebner blamed the fiasco on "contaminated fuel," a face-saving excuse that likely covered up some kind of mechanical failure.[32]

Despite this minor glitch, the publicity campaign generated lots of favorable press for Chrysler, garnering articles in major newspapers and national magazines—from *Time* to the *Saturday Evening Post*. The release of fifty Turbines to ordinary people was framed as an "experiment" in marketing, putting the car through its paces under a variety of realistic driving conditions. Chrysler could then gauge consumer reactions before deciding on mass production.

Chrysler solicited applications from people across the country who wanted to receive free use of a Turbine, including insurance and repairs; the only expense would be fuel. From the thirty thousand or so people who applied, an outside auditing firm selected 203 to drive the Turbine for a period of three months each. The lucky users lived in every state except Alaska and Hawaii. My impression is that most users were middle and upper-middle class; about 90 percent of the drivers were men.[33]

The selection of each family was accompanied by a barrage of Chrysler-orchestrated publicity in local media. The family took its first ride in the car as part of a press event. Actually, the drivers had become familiar with the car beforehand—Chrysler was taking no chances.[34]

Richard and Patricia Vlaha of Chicago were the first people to receive the car. They learned, as did all users, that the car was a people magnet. Wherever they stopped onlookers gathered and asked question after question. Steve Lehto, who wrote the definitive book on the Turbine, managed to interview Richard Vlaha and several other lucky users. The Vlahas "loved the car" but it had "horrible mileage" in the city and not much more than ten miles per gallon on the highway, similar to the worst gas hogs then on the road. Although Chrysler claimed far better mileage for the car, the company was so concerned that it forbade Vlaha (and perhaps other users) from talking about fuel economy.[35] But this was the 1960s, a decade before the first OPEC oil embargo when fuel—anything but leaded gasoline could be burned in the Turbine—was still cheap. Even so, a car's fuel economy was an owner's point of pride, frequently bragged about but more often lied about.

In states with extremely cold winters, Turbine users reported that the car started with no difficulty and (best of all) that the heater instantly pumped out very warm air. However, starting required a 24-volt battery. Simple repairs were done in the field at no cost to the user, but any internal engine problem led to its complete replacement—a new engine arriving by air freight. A Chrysler technician who serviced the cars told Lehto that the engine was probably replaced once in every car. Whenever a general problem was found, as in the starter, the entire Turbine fleet received the repair or upgrade during the few weeks it spent between users.[36]

The car had some quirks. Chrysler instructed drivers to wait until the tachometer reached twenty thousand revolutions per minute before putting the car in gear. Impatient drivers ignored this instruction and as a result badly damaged the engine seals, causing a loss of power. The engines had to be replaced, but a fix was soon installed that prevented the engine from going into gear until it had revved up. Some users also damaged the automatic transmission by shifting it manually at the wrong speed.[37] If a car ran out of fuel, the user was unable to start it and had to telephone Chrysler to send a technician.

Throughout the two years that the Turbine was on the road, Chrysler orchestrated many publicity events—even at shopping centers—that kept the car in the public eye. Was the Turbine a creative and expensive marketing gimmick? One might draw this conclusion and not be far wrong. In a film aimed at Chrysler dealers, an executive claimed that "the campaign would generate two years of public relations that showed Chrysler on the cutting edge of technology."[38] But to the engineers, especially Huebner, it represented more than a dozen years of hard work as they attempted to jumpstart an entirely new automotive technology. New technologies are often a public relations bonanza for a company as well as a significant technical accomplishment, and the Turbine was both. Beginning in 1962 Chrysler's U.S. market share began to increase appreciably. The sixties would be a profitable decade for Chrysler; perhaps the shiny bronze cars helped spur sales of its products.[39]

After the last car was turned in on January 28, 1966, the Turbine loan program ended but the publicity continued. Turbines toured college campuses, where Chrysler hoped they might entice engineering students to apply for employment. As an undergraduate at UCLA, I witnessed one of these show-and-tells, with the Chrysler car idling in front of the engineering building.

In a press conference on April 12, 1966, Chrysler announced the findings of the Turbine test program. Huebner and David F. Miller, Chrysler's manager of marketing and consumer research, read prepared statements. Huebner covered the technical details at some length, acknowledging many problems and their fixes, but also claiming that "our inspections of high-mileage cars indicate that most of those parts will enjoy far longer life than 50,000 miles and pass the 100,000-mile mark." He also insisted that "no unusual problems" would be encountered in training people to maintain and repair the engines. Huebner emphasized that the Turbine's engine "was not to be considered a final production design," but that "its basic concept" was sound and had the potential for mass production.[40]

Within two weeks after each Turbine was returned, members of Chrysler's marketing staff conducted an intensive personal interview with the user. Miller reported that "users generally were exceptionally enthusiastic about the turbine car." They listed the turbine engine's main advantages as "vibrationless, smooth operation" and "reduced maintenance requirements."[41] They were also impressed with the car's ability to start rapidly in any temperature and produce heat instantly. I suspect that these favorable performance characteristics alone were not responsible for the user's enthusiasm. We must keep in mind that every time these cars stopped, people nearby approached the car and asked the driver questions. This put the driver in the role of expert as he or she explained the workings of the unusual car. While perhaps a nuisance at some level, such experiences also conferred prestige and enhanced the user's social power, especially among friends and relatives. Perhaps the Turbine's social effects might have strongly biased the users' positive overall evaluations.

The interviewers also learned about the Turbine's disadvantages. One-third of the people were displeased with the accelerator lag, which had been reduced to about 1.5–2.0 seconds, and one-quarter were disappointed with the fuel economy. Some also complained about the difficulty of finding unleaded gasoline and diesel fuel. Miller put an upbeat spin on the findings, concluding that "the advantages of turbine power are readily apparent to typical motorists. The consumer evaluation has uncovered some areas of performance that require improvement. However, if these shortcomings can be overcome, it is evident . . . that the idea of turbine-powered passenger cars is capable of earning widespread consumer acceptance."[42] That is, if and when a turbine engine performed as well as a piston engine, people would buy turbojet cars.

At the same press conference, Harry E. Chesebrough—vice president for product planning and development—promised that Chrysler would make a decision by early summer "about moving into the next phase of our turbine program." In view of the statements by Huebner and Miller, which indicated that the turbine engine still needed work, Chesebrough implied that the turbine program might be at risk.

In September, after what must have been much soul-searching by Chrysler executives, Chesebrough reported that "Chrysler Corporation is going ahead with development of a new, fifth turbine engine." But inquiring minds wanted to know whether Chrysler would produce a second generation of the Turbine and bring it to market. Curiously, according to an article in the *New York Times*, Huebner "thinks the engine is ready for the street, but so far his corporate bosses say no."[43]

Chesebrough was explicit about what conditions had to be met before Chrysler would begin making turbojet cars for the masses. "When we are satisfied we have a turbine design that is capable of being mass produced at a suitable economic level, as well as being at least the equal of the piston engine in performance, fuel economy and reliability, we will freeze the design and begin the complex process of developing the tools and facilities necessary for mass producing the engine." How long this might take, he could not predict. Putting the best face on the disappointing news, he hoped that the company's "efforts will be successful so that Chrysler Corporation will be able to offer a revolutionary, improved power plant to American motorists."[44]

Development of the fifth-generation engine was already well along. It would be more powerful so that a car could have air conditioning. In subsequent years Huebner and the Chrysler team continued turbine development. The sixth-generation engine had reduced accelerator lag and greater power. During the 1970s, with a $6.4 million contract from the Environmental Protection Agency, they worked on the seventh-generation engine. Some progress was made but problems remained.[45] In any event, for many reasons Chrysler never mass-produced a turbine-powered car. Thanks to government contracts, however, Chrysler's expertise in turbojet engines did not go to waste. In 1978 the company began manufacturing the Abrams M-1 tank whose 1,500-horsepower turbine engine was very reliable, but the vehicle recorded less than one mile per gallon.[46]

Chrysler was not alone in tackling the turbine. Virtually every major automaker in the United States, England, Japan, and Germany invested in turbine engine development, but none brought a jet-powered car to market. While these engine projects were under way in the 1950s and sixties, citizens were clamoring for cleaner air. For example, in Los Angeles—where I grew up—the smog was so bad on many days that our eyes burned and lungs hurt and visibility was reduced to a few blocks. We learned later that cars were the major source of air pollution, pumping out carbon monoxide, hydrocarbons, and nitrogen oxides (this was before the greenhouse gas, carbon dioxide, was of concern).

To reduce the amount of gaseous pollutants entering the air, Congress phased in rigorous standards for all new cars to take effect beginning in 1975.[47] Because fuel is burned so completely in turbine engines, they produce little carbon monoxide and few hydrocarbons, but the high combustion temperatures increase the yield of nitrogen oxides—which contribute greatly to smog.[48] Lowering nitrogen emissions by tweaking the engine's design would have required painful compromises in other performance characteristics. Conventional cars, which also produce nitrogen oxides, were eventually outfitted with catalytic converters.

Adding to the turbine engineer's difficulties were fuel economy standards enacted by Congress: by 1985 a company's fleet had to average 27.5 miles per gallon. American carmakers complained loudly and lobbied hard, claiming that the standards were technically impossible to meet. However, an increasing influx of durable, fuel-efficient compact cars from across the seas—including Toyotas, Hondas, and Volkswagens—exposed this claim as self-serving blather. American companies at last downsized their cars and designed more fuel-efficient engines using electronic ignition and fuel injection. Even so, they continued to make thousands of highly profitable gas-guzzlers whose poor fuel economy was offset by the compacts and subcompacts in their fleets—some of them made by Japanese partners.

Chrysler could have installed turbines in a line of luxury cars and produced them in small numbers. Such cars might have been attractive to some consumers owing to their novelty and high-tech cachet. The company explored this possibility but gave up after an underpowered Chrysler Imperial performed poorly.[49] How many affluent customers would pay extra for the bragging rights afforded by ownership of a jet-powered car? No one knew.

Many people blame federal regulations for killing the turbine car, but there were also significant production-related barriers. No matter the scale of mass production—ten thousand or a hundred thousand cars per year—building the engine in quantity would be a major challenge, for it was devilishly difficult to manufacture. Investing in a new factory and developing new production equipment and processes would be an immensely expensive commitment. Also, a new workforce with special training would be needed. And service departments in dealerships would have to keep an inventory of replacement parts and ensure that their technicians had the expertise to work on turbines.

Forecasting whether these investments, perhaps hundreds of millions of dollars, were likely to pay off would be a tricky exercise in cost accounting. Clearly, no car company in the world—even the most prosperous—calculated that this was a risk worth taking, especially not Chrysler. The company had enjoyed robust sales and profits during the 1960s but the situation had changed for the worse in the 1970s and 1980s. To a company under increasing financial pressures, as was Chrysler, it would have been unthinkable to mass-produce a new car as risky and expensive as a second-generation Turbine.

Finally, there was the matter of competing with piston engines. Turbine engine advocates misjudged the piston engine's potential for further development, as in Robert C. Farmer's 1971 claim that the conventional engine "stands little chance of remaining a major source of vehicle power, and still satisfying government regulations and customer needs."[50] From the 1960s through the 1980s, however, cars with piston engines—compact foreign cars and downsized domestic ones—exhibited vast and continuous improvement in fuel economy and reliability and emitted fewer pollutants. Turbine engines did not compete well against this moving target. As even Huebner tellingly admitted, an alternative power plant would have to be "superior in every way over current engines."[51] Even if the turbine engine's performance and production problems could have been solved, it is doubtful that it would have performed better than the piston engine in all respects.

From time to time since 1980, companies around the world have had eruptions of interest in developing improved turbine engines for passenger cars, often taking advantage of government funding. There have also been strong claims of progress.[52] Conforming to an old pattern, companies use turbine-powered test cars to advertise that their engineers are hard at work on the cutting edge of a revolutionary technology.[53] In 1981, for example, Mercedes–Benz (with funding from the German government) publicized its creation of an engine using several silicon nitride and silicon carbide parts. These ceramic components permitted turbine inlet temperatures exceeding 2200°F.[54] Ceramic components improve efficiency but their use presents durability problems, for they are brittle materials.[55] The larger problem is that incremental engine improvements, while enhancing a company's public image, are unlikely to change its calculation of the risks and benefits of mass-producing a turbojet car.

Huebner, of course, never gave up on the turbine engine. As late as 1979, he claimed that his choice for an alternative to the piston engine remained the turbine. "I believe my reasons for choosing the turbine are objective and that they are based on sound scientific principles."[56] But scientific principles alone motivate neither consumers nor corporate executives.

✳ ✳ ✳ ✳ ✳

What happened to the Chrysler Turbines? Late in 1966 the company destroyed all but nine of these handsome automobiles: two remain with Chrysler; five are in museums, including the National Museum of American History, Smithsonian Institution; and two are in private hands, one owned by Jay Leno who bought it from Chrysler and enjoys driving it around southern California.[57] As of 2018, however, no ordinary consumer can purchase a turbojet passenger car from any company.

NOTES

[1] I thank Roger White, curator of transportation at the National Museum of American History, Smithsonian Institution, for giving me access to materials on the Chrysler Turbine.

[2] Information in this paragraph comes from Constant (1973).

[3] https://en.wikipedia.org/wiki/North_American_F-86_Sabre#Korean_War, accessed 21 March 2016.

[4] The comprehensive source on early turbine car developments is Norbye (1975).

[5] On Rover's program see Norbye (1975:237–264).

[6] Norbye (1975:319–393) discusses General Motors' turbojet work.

[7] On Chrysler's innovations compared with those of Ford and General Motors, see White (1971:302, Table 13.1).

[8] Langworth and Norbye (1985:69).

[9] Braun (1992b) summarizes Chrysler's efforts.

[10] Braun (1992b:341–342).

[11] Huebner (1964/1965:2).

[12] Norbye (1975:265–318) supplies technical details on generations 1–6.

[13] Ludvigsen (1961); "Chrysler Tries Out Gas Turbine Engine." *New York Times*, 25 March 1954, p. 52.

[14] Lehto (2010:12).

[15] Lehto (2010:11).

[16] Huebner (1964/1965:5).

[17] Norbye (1975:274).

[18] "Gas Turbine Powered Car Arrives in L.A." *Los Angeles Times*, 31 March 1956, p. A1.

[19] "Gas Turbine Powered Car Arrives in L.A." *Los Angeles Times*, 31 March 1956, p. A1.

[20] Huebner (1964/1965:5). Use of the term "gas" distinguishes the car turbine from water and steam turbines.

[21] In later years some compositional information was published (e.g., Roy, Hagen, and Belleau 1964).

[22] Huebner (1964/1965:8).

[23] Huebner (1964/1965:15–16).

[24] Huebner (1964/1965:16).

[25] Huebner (1964/1965:17).

[26] Huebner (1964/1965:17).

[27] Unless otherwise noted, information on Chrysler's Turbine comes from Lehto (2010).

[28] Lehto (2010:40).

[29] Description of the Turbine in the following paragraphs comes from Lehto (2010:42–46).

[30] The transmission lacked a torque converter because the engine itself served as the torque converter; the engine's gas generator and power turbine sections were not connected by a shaft.

[31] For an informative video on Jay Leno's Turbine, see https://www.youtube.com/watch?v=b2A5ijU3Ivs, accessed 8 July 2018.

[32] Lehto (2010:49–50).

[33] Haitch (1956).

[34] Lehto (2010:54–55).

[35] Lehto (2010:58–59).

[36] Lehto (2010: Chapter 8).

[37] Huebner (1966).

[38] Quoted in Lehto (2010:68).

[39] White (1971:251, Table 15.2).

[40] Quotes in this paragraph are from Huebner (1966).

[41] Miller (1966).

[42] Quotes in this paragraph are from Miller (1966); some information comes from Chrysler Corporation (1966).

[43] Flint (1967).

[44] Chesebrough (1966).

[45] Farmer (1971).

[46] http://www.allpar.com/history/military/M1-tanks.html, accessed 6 April 2016.

[47] Kaiser (2003).

[48] Braun (1992b).

[49] Lehto (2010:130).

[50] Farmer (1971:14).

[51] Huebner (1979:3).

[52] Harmon (1982).

[53] Banovsky (2015).

[54] Norbye (1981).

[55] Katz (1989:10–15); McLean (1970).

[56] Huebner (1979:12).

[57] http://www.allpar.com/mopar/turbine.html, accessed 5 April 2016.

12

THE CONCORDE

Supersonic Airliner

Crossing the Atlantic Ocean has always been a trial, but during the past two centuries new transportation technologies have made travel more tolerable. A trip that took Benjamin Franklin eight weeks in a sailing ship now takes eight hours in a jumbo jet. Gains in speed, however, have not always been matched by gains in comfort. Today's eight-hour trip still has cramped passengers literally aching for a faster flight.

In the immediate postwar period, when the first nonstop trans-Atlantic air travel had begun with piston-engine planes and long-range jets were still some years away, aviation visionaries were inspired by cutting-edge military aircraft. After all, during the mid-1950s jet-powered fighter planes in England, France, the United States, and the Soviet Union were already passing the sound barrier in sustained, level flight (Mach 1 = 761 miles per hour at sea level).[1]

In view of the long-term trend toward faster aircraft, it seemed reasonable to forecast that supersonic flight would also be the next step in civil aviation. Yet the challenge to creating supersonic airliners was not in propelling a plane past Mach 1, but in devising a comprehensive design that was—at a minimum—safe and reliable. A further issue was whether such an aircraft could be economically feasible, given that its total development time and costs were unknown at first and that the potential market was murky at best.

219

✳ ✳ ✳ ✳ ✳

A project to build a supersonic airliner had been germinating in Britain since the early 1950s, even as de Havilland launched the Comet—the world's first commercial jetliner—in 1952. The British Comet had severe design defects and was grounded after four crashes.[2] Across the Atlantic Boeing was developing its first passenger jet, the durable 707, and Douglas was developing the DC-8. Both American manufacturers were building large-capacity planes for long flights that by 1960 would help democratize air travel for the growing middle class.

Owing to the reborn Comet's dismal sales, Britain had every reason to fear that American firms would soon dominate the world market for subsonic airliners. Perhaps a supersonic airliner would be the last chance for Britain's aviation industry to assert world leadership. Creating such a plane to leapfrog the new American jets would be an expensive and risky undertaking, requiring formidable engineering and scientific achievements, but it would also be a potent symbol of the country's postwar resurgence.[3]

Although the British government had already begun contracting for supersonic-related research, including one-off aircraft to test various concepts, it was now time to launch a formal program. In November 1956 Britain's Supersonic Transport Aircraft Committee (STAC) held its first meeting in London. The meeting was led by aeronautical engineer Morien Bedford Morgan, deputy director of the Royal Aircraft Establishment at Farnborough, who would become known in England as the "father of the Concorde." In attendance were representatives from aircraft and engine manufacturers, research establishments, the state-owned airlines, and several government ministries.[4] In the next few years, STAC subcommittees attempted to show that a supersonic airliner was technically and economically feasible. The discussions covered seemingly every topic relevant to the plane's design, from wing shape to handling qualities to propulsion system, and even touched on social impacts such as sonic booms and noisy takeoffs and landings.

Financed by government ministries, British aviation companies worked on two designs: a short-haul 100-seat plane capable of

Mach 1.2, and a long-haul 150-seat plane that would cruise at Mach 1.8. In March 1959 a STAC report, accompanied by five hundred supporting documents, summarized the research findings and urged pursuit of both options.[5] Although the report did not focus on the economics of a supersonic transport plane (SST), and airline representatives worried about the size of the market, Morgan's position was well known. He believed that an SST would advertise the country's technical skill and "might be a good investment for Britain as a whole, even it if lost money."[6] Evidently, as a symbol of technical prowess an SST would be a political technology also having concrete benefits, such as keeping aircraft companies afloat and employing legions of skilled workers. And an SST might even carry people swiftly and safely across the Atlantic.

Britain sought a partner to share the costs and risks of developing an SST. Protracted negotiations between British and American government officials, as well as between aviation companies in both countries, did not result in an agreement.[7] Apparently, the United States wanted to go it alone. However, building on earlier Anglo-French collaborations between their aviation industries that had produced the Caravelle airliner and the capable Mirage III fighter, France agreed to partner with Britain.[8] A successful SST would no doubt provide ample prestige for both countries.

In November 1962 Julian Amery, the British minister of aviation, and French Ambassador Geoffroy de Courcel signed an agreement on behalf of their governments to jointly develop and manufacture an SST. The project would be financed by government contracts to aircraft manufacturers in both countries. France was eager to move ahead, but British skeptics kept pointing to the project's vast uncertainties. Although no one could know how many SSTs would be sold, Morgan and STAC forecast that 150–500 SSTs would be flying and revolutionizing long-distance travel.[9] Making the case for an SST before the House of Commons, Amery claimed that "this aircraft has every chance of securing a substantial part of the world market for supersonic airliners. This is a chance that will not return."[10] Once the agreement was signed, the British and French governments moved ahead. Even naysayers in London couldn't derail the project, which had become a national priority. After delicate negotiations the SST was dubbed the Concorde.

Throughout the arduous development process, policymakers and engineers in both countries laid stress on solving every conceivable problem and rigorously testing the solutions. The Concorde could be nothing less than a brilliant technological success. This meant no shortcuts, no risky workarounds; virtually every decision would require across-the-board agreement. And so the development process moved ahead at a deliberate—some might say plodding—pace.

The engine selected for the Concorde (each plane required four) was the Rolls-Royce/Snecma Olympus 593. Although the design of jet engines had matured greatly by the mid-1950s, perfecting this power plant to meet the performance requirements of an SST was a difficult undertaking.[11] This engine would power all Concorde aircraft, eventually becoming renowned for "its longevity, reliability and power" and high thermal efficiency.[12]

Beyond the engine, the Concorde presented an array of challenging engineering problems. To allow a plane to move rapidly through the air with little resistance, the fuselage needed to be very slender, as in a small-diameter tube. Too small, however, and passenger seating would be limited. A compromise was reached giving the Concorde two seats on each side of a central aisle. Eventually, the two-plane plan was abandoned and a hybrid design configured. The Concorde would seat a hundred passengers and fly at Mach 2 with a range of 3,500 miles. This configuration would permit the plane to cross the Atlantic in less than four hours, reducing travel time between New York and London by more than half.

A thin wing was another aerodynamic necessity, but it would furnish insufficient lift on existing runways, making takeoffs and landings somewhat problematic. Engineers could not expect airports to extend their runways by half a mile or more just to accommodate the Concorde. In designing a wing that would be thin with ample lift, engineers experimented with hundreds of models informed by "mathematics, fluid dynamics, aerodynamics, imagination and, of course, trial and error."[13] The resultant wing required a somewhat steep approach, in both takeoffs and landings, to form vortices—miniature tornadoes—on the wing's upper surface, which effectively sucked the plane upward.[14]

To learn whether the thin wing would work as well as predicted at low speeds, the British company Handley Page was awarded government contracts to build the HP.115. It had a very slender fuselage and a swept-back, delta wing. Veteran test pilot Brian Trubshaw, who also put in exemplary service on the Concorde, described the plane as "a glider with a small engine."[15] Flying at very low speeds, the HP.115 did not stall and supplied much useful data.[16]

In landing at such a high angle, the pilots could not see the runway, their vision confined skyward. The solution was the Concorde's droop nose, which debuted on Fairey Aviation Company's Fairey Delta 2. The plane was designed to test high-speed flight as well as takeoffs and landings. In an early flight, the Fairey Delta 2 cruised at 1,132 mph, setting a new speed record and validating predictions about high-speed stability.[17] Approaching a landing, the plane's nose gently swung down, allowing pilots to see the ground. Today, of course, that problem can be solved with digital electronics. But this was the 1950s when video cameras were huge and heavy. An inelegant solution, the droop nose did work and gave the airliner its distinctive appearance, which Concorde historian Jonathan Glancey likened to "a magnificent mechanical stork."[18]

Choosing a material for the airframe—possibly stainless steel, aluminum, or titanium—required tradeoffs among strength, weight, and cost. A Bristol 188 was built to evaluate the use of stainless steel. In what seems like a painfully obvious finding, the Bristol 188 showed that a stainless steel plane requiring eight engines was far too heavy. After many tests an aluminum-copper alloy was selected for the Concorde's airframe.[19]

Despite impressive efforts of Concorde engineers, two problems resisted solution. A sonic boom was inevitable as the plane passed through the sound barrier, unleashing a deafening noise that spread to an area 25 miles wide along the supersonic flight path.[20] Sonic booms were annoying and nerve-wracking, but the bomb-like concussion could also damage structures. In a 1964 test that seems unfathomable to us today, Oklahoma City was subjected to eight sonic booms per day for six months. The result was "15,452 complaints and 490 claims for damages, mostly for shattered windows."[21]

Because the Concorde had to take off and land at relatively high speeds, the roaring engines would be a problem around urban airports. In an ominous move, some cities—including Los Angeles—imposed noise limits that ruled out SST service. Britain and France were outraged that a locality could make a law regulating international air traffic. Was the United States trying to hobble the Concorde, perhaps to prepare the way for an American-built SST? In the long march to the Concorde, rivalries and jealousies simmered just below the surface of international relations, sometimes abetted by spying.[22]

Somewhat insulated from the international intrigues, the engineers continued to refine the Concorde's design, giving special attention to the craft's overall weight. Yet many redesigned features added pounds, leading to a takeoff weight of four hundred thousand pounds—half of which was fuel. To shave weight, engineers had to adopt more expensive construction methods and more expensive materials. According to Peter Baldwin, permanent secretary at the British Department of Transport, the "cost turned out to be about four times as high as had been anticipated. The economics . . . had never been very good, but it became obvious from the mid-1960s to anybody that this was economically disastrous."[23] As development costs soared, debates in Parliament raised numerous questions about the advisability of continuing the project. But there was no escape: the 1962 agreement "lacked a provision for unilateral withdrawal."[24] In France, completing the project at almost any cost had become a matter of national pride.

The Concorde was manufactured jointly by the British Aircraft Corporation and Sud-Aviation (later Aérospatiale) of France (Figure 12.1). Assembly entailed many logistical problems because the factories—one at Filton, near Bristol; the other at Toulouse, near the Spanish border—were six hundred air miles apart and parts manufacturers were dispersed even more widely. This required the constant transport of people and parts, which increased production costs.[25] In the end each plane cost more than $150 million to produce, on top of nearly $3 billion spent on development.[26]

Beyond development and production costs, an important indicator of a plane's economic viability is the percentage of its total

Figure 12.1. Concorde's last flight, 26 November 2003, overflying Filton Airfield. Source: courtesy of Wikimedia Commons, photographer Arpingstone.

weight dedicated to payload (people plus luggage and cargo). The higher the percentage, the better—because payload pays. Unable to carry cargo, the Concorde's number was 6 percent, half that of a Boeing 707. And by the time the Concorde was flying, jumbo jets had exceeded 20 percent.[27] Six percent was not a showstopper but it did portend high operating costs, a serious concern for any airline. Despite the many red flags waving insistently that the Concorde might become the white elephant of the sky, the project lumbered on.

Beyond formally setting the project in motion, the Anglo-French agreement of 1962 had two additional effects. First, it generated 36 orders for the plane by the end of the year. An airline's "order" was merely an option to buy, not an ironclad contract; even so, this was a respectable number for an aircraft that did not yet exist. Airlines reasoned that if the Concorde delivered on its performance promises, every major carrier would require supersonic routes in order to remain competitive. Whether that interest would translate into actual sales, however, remained an open question into the 1970s.

A second effect of the 1962 agreement was to bring into the open a race for supersonic supremacy that included the Soviet Union and the United States. If supersonic flight were to be the next step in civil aviation, then the two superpowers would need their own SST projects, an imperative of Cold War high-tech competition. And, as in Britain and France, the Soviet and American SST projects would depend on generous government funding.

During a commencement address at the U.S. Air Force Academy on June 5, 1963, President John F. Kennedy made a surprising announcement. The United States, partnering with private industry, would create an SST program "to develop . . . a commercially successful supersonic transport superior to that being built in any other country." Tellingly, he emphasized that "neither the economics nor the politics of international air competition permits us to stand still in this area."[28]

Advocates in the administration argued that without an American-made SST, airlines would buy the Concorde, sending billions of dollars abroad. In addition, the country would lose the opportunity to create tens of thousands of well-paying jobs.[29] Critics in Congress—especially Senator William Proxmire—panned the project, arguing that an SST would pollute the atmosphere, that the projected sales figures were vastly inflated, and that private enterprise should pay for it. The project went ahead anyway.[30]

The U.S. government would supply 90 percent of the development costs, the rest to be divided among the winners of the final competition. This rather one-sided public–private partnership eventually cost the U.S. government "more than $1 billion."[31]

Following a military procurement model, the SST program began with a design competition among American aviation companies. At the end of 1966, winners of the final design competition were announced. Some observers had expected Lockheed, which built the SR-71 "blackbird" (the Mach-3 spy plane), to be the winner. Instead Boeing and General Electric were chosen. Boeing, which had been working on supersonic designs since 1952, would build the airframe; General Electric, experienced maker of turbojet engines, would build the power plant.

Boeing's original design for a partially retractable wing turned out to be unworkable. The revised design, worked out in great detail, presented a plane that in principle would leave the Concorde in the clouds. Not only would the Boeing 2707-300 be the fastest SST, cruising at Mach 2.7 (1,800 mph) on four GE-4 engines, but it would also be the largest by far—able to carry anywhere from 253 to 321 passengers for a distance of four thousand miles. From droop nose to tail, the titanium airframe would measure 298 feet. Even with a takeoff weight of 750,000 pounds, the SST would fly supersonically at 62,000 feet, able to go from New York to Paris in a little less than three hours. To materialize their design, Boeing built and publicized a full-scale mock-up of the 2707-300.[32]

Boeing insisted that the plane would be economical to fly, claiming that "the total operating costs and attractiveness of its speed and ride comfort to the passenger will combine to achieve reasonable airline profits at today's economy class ticket prices."[33] On the basis of such unfounded promises, by the end of 1969, 26 airlines— with their eyes on competitors—had placed orders for 122 planes; the future looked bright for the Boeing SST. The optimistic projections for total aircraft sales and operating costs assumed that fuel prices would remain at historic lows.

In the meantime Boeing's plane ran into stiff headwinds. Environmental organizations and others raised issues about damage to the ozone layer, airport noise, and the effects of sonic booms. These issues resonated widely, leading to a 1973 federal ban on supersonic flights over land in the United States that would also affect Concorde by prohibiting potentially lucrative cross-country routes.[34] Although President Nixon strongly favored continuing work on the SST, some in his administration were opposed. Responding to mixed messages and contradictory forecasts of the plane's economic prospects, Proxmire and his allies in Congress were able to kill federal funding for the SST in the spring of 1971.[35] Boeing and General Electric did not continue the project on their own, despite their rosy forecasts of the plane's operating costs and the many advance orders.

✳ ✳ ✳ ✳ ✳

The Soviet Union, desperate to prove that its totalitarian state could deliver technological wonders superior to those of the Western democracies, formally began development of its own SST in July 1963.[36] Although starting late, the Soviet Union aimed to upstage the Anglo-French alliance by flying first and, eventually, securing orders from airlines around the world. The Tupolev Design Bureau, a government agency, was in charge of the rush project to build an SST, named the Tu-144. The plane was to have a capacity of 120–135 passengers, flying slightly above Mach-2 with a range of four thousand miles.[37]

Figure 12.2. Russia's supersonic Tu-144LL taking off at Zhukovsky Air Development Center (1997). Source: courtesy of NASA.

The Tu-144 with its droop nose so greatly resembled the Concorde that it was known derisively in the West as "Concordski" (Figure 12.2). The similarities were superficial, however, and the differences—in engine performance, passenger comfort, and reliability—were significant. Even so, the Tu-144's first flight on December 31, 1968, with test pilot Eduard V. Yelyan at the controls, beat Concorde to the air by two months. The plane, Yelyan claimed, was easy to fly.[38]

Unlike the Concorde, whose prototypes went through 5,495 hours of flight testing, the Tu-144 prototype was tested for only 800 hours.[39] A trial period this brief hinted that problems might arise in actual service—and they did. Even before the plane carried its first passengers, the project suffered a serious setback in June 1973 when a Tu-144 crashed at the Paris Air Show, killing fourteen people. According to the *New York Times*, the crash dealt a blow to Soviet ambitions "to corner at least part of the world's supersonic airline market."[40] In the meantime this bitter humiliation indicated a pressing need for further development, thus delaying the start of passenger service.

At last, in November 1977 nine Tu-144s began carrying passengers on Aeroflot, the state-owned airline, but substantial operating costs and poor reliability were serious problems. After the crash of an "improved" version and only 55 scheduled flights, the Tu-144 was permanently grounded as a passenger plane. However, the plane continued to carry cargo and mail and served as a test bed for experiments by the Soviet space program—and, ironically, by NASA. The accelerated development program had taken its toll. "[T]he Tu-144s suffered 226 failures, eighty in flight, meaning delays, cancellations and heart-stopping diversions."[41]

Although having been beaten to the air by the TU-144, the Concorde amassed one milestone after another in hundreds of test flights. Departing Toulouse, Concorde 001 first flew on March 2, 1969, with French test pilot André Turcat at the controls; in October Turcat pushed the plane past Mach 1.[42] Brian Trubshaw, who first flew Concorde 002 in England, was pleased with the Concorde's "excellent and precise handling qualities."[43] No doubt this World War II bomber pilot was also comfortable with the plane's analog gauges and controls. During more than five years of rigorous test flights, Concorde's go-slow design process was thoroughly validated. The plane was certified airworthy in late 1975.

In 1973, in the midst of the Concorde's lengthy testing phase, the Organization of Petroleum Exporting Countries (OPEC) unleashed its first oil embargo—causing an energy crisis. Long lines at gas pumps underscored the heavy dependence of the West on petroleum from the Middle East's feudal monarchies. The fourfold increase in the price of aviation fuel (kerosene) augured poorly for the Concorde's commercial prospects. After all, the aircraft's fuel consumption per passenger mile was three times higher than that of the new jumbo jets.[44] Also threatening the Concorde was increasingly restrictive legislation in many countries and localities. Given these obstacles, would the international airlines turn their early options into concrete purchases?

Although the British and French governments could afford to pour billions into developing political technologies, airlines had to carefully weigh the prestige of supersonic capabilities against the plane's rising purchase price and steep operating costs. The Concorde's estimated price in the 1960s had been less than $20 million, but in late 1971 the plane was listed at $31.2 million and eventually reached $60 million.[45] In the meantime the 400-seat Boeing 747 and McDonnell Douglas DC-10 jumbo jets—priced at $23 million and $15 million, respectively—were becoming ubiquitous on international routes, carrying record numbers of passengers for fares that middle-class people could afford.

Not surprisingly, airlines began to reassess their plans to buy supersonic aircraft. Pan Am and TWA had long been industry bellwethers, often the first adopters of new planes; together they held options on thirteen Concorde aircraft. In early 1973, even before the first OPEC oil shock, both airlines canceled their orders. Pan Am "ascribed the move to the plane's significant inferiority, compared to jumbo jets, in range, carrying capacity and costs."[46] These cancellations, admitted French Transport Minister Robert Galley, were a "hard blow."[47] And then it got worse. By the end of 1975, before a single paying passenger had flown on a Concorde, *all* other airlines had dropped their options.[48] The financial debacle long predicted for the Anglo-French SST had come to pass, but both countries soldiered on with the project.

The Concorde's first—and only—buyers were the national airlines of Britain and France, which eventually acquired seven planes each, the last few at no cost. Twenty years after the first meeting of STAC, these two carriers launched passenger service early in 1976: British Airways, London to Bahrain; Air France, Paris to Rio de Janeiro with a stop in Senegal. Scheduled flights were added to Washington, D.C., in 1976. After lengthy legal wrangling that reached the U.S. Supreme Court, in 1977 the Concorde at last gained the right to fly from Paris and London to New York.[49]

Unlike the privately funded jumbo jets that served middle-class travelers, the Concorde was financed by the taxes of ordinary French and British citizens, yet it served only a niche market of elite consumers. In 1994 the New York–London round-trip ticket cost $8,756—gourmet meal and champagne included. The trans-Atlantic flights catered to "admirals of industry, senior politicians, diplomats and civil servants."[50] An enticement to the business community was the possibility of leaving London at 8:00 A.M., arriving in New York at 7:00 A.M., doing a partial day's work, and returning home by midnight. Although businesspeople were the main travelers, the novelty and convenience of supersonic flight also attracted wealthy celebrities of every kind—from Elizabeth Taylor to Queen Elizabeth II—and musicians by the score including John Lennon, Elton John, Mick Jagger, Michael Jackson, and Paul McCartney. Flying on the Concorde had become a "must do" for the rich and famous. Nonetheless, many seats remained empty and British Air resorted to charter flights to ramp up revenue.

Elite clientele and charters kept the Concorde in the air as long as British Airways and Air France were willing to absorb the mounting financial losses. In 1977–1978 British Airways lost $32.6 million on Concorde routes, while in 1979 Air France lost $10 million.[51] With no further aircraft sales imminent despite round-the-world promotional tours, production of the Concorde was ended in 1979. In both countries tens of thousands of workers lost their jobs. To further stanch the financial bleeding, routes were cut. By the end of 1994, only scheduled service from London and Paris to

New York remained. Through the white magic of cost accounting, these routes were claimed to be profitable.

The Concorde flew uneventfully for more than two decades, amassing almost two hundred thousand hours in the air.[52] Everything changed suddenly on July 25, 2000, when a freak accident caused Air France Flight 4590 to crash on takeoff from Paris, killing 113 people. British Airways continued flying after this disaster, but Air France suspended service for more than a year while determining the cause of the crash. Although the Concorde's safety record was marred, conservative longevity estimates had the aircraft flying for five or ten more years.[53] However, increasingly oppressive maintenance costs and dwindling ridership led both airlines to retire the entire fleet in 2003. And so ended supersonic passenger travel, precisely a century after the Wright brothers proved the feasibility of controlled, powered flight.

The Concorde was a stunning technological success that deserved its many accolades. It was, after all, "an aircraft that could fly the Atlantic in three hours at twice the speed of sound without so much as rippling the surface of a gin and tonic."[54] But the Concorde was also a dismal commercial failure, bought only by the national airlines of Britain and France. Let us recall that at the outset many people judged the Concorde to have poor economic prospects. Even so, in both countries the project moved ahead at considerable public expense. Clearly, unfavorable economic calculations were discounted because this sophisticated technology's primary function was largely symbolic.

The Concorde was a political technology that successfully touted the industrial capabilities of Britain and France to the world. After the Tu-144's short-lived passenger service ended in 1977, only the Concorde ferried passengers supersonically. Even the United States failed to achieve an SST, the federal government having decided to cut its losses before beginning production. The Concorde also showcased a successful international collaboration, a pattern that has been followed to an even greater extent by Airbus—a fierce competitor to Boeing financed by a consortium of European governments.

There is no mystery as to why all air carriers, except the captive national airlines, shunned the Concorde. Expensive to buy and expensive to operate, the Concorde was forecast to be a money pit. Adding to the obstacles were the restrictions on overland flights and noise limits at many airports, which combined to reduce the available routes. From the perspective of every airline, the Concorde was not expected to be a profitable investment.

The vision of supersonic airliners lives on, seemingly a cultural imperative awaiting realization.[55] During the 1990s Boeing did a study for NASA that assessed the potential for creating a new SST. Optimistic projections indicated a need for supersonic fleets between the years 2000 and 2005. In wide-ranging analyses covering "technology, economics, and environmental constraints," Boeing reprised many features of the defunct 2707-300. The study's conclusion was sobering. "[C]urrent technology cannot produce a viable high-speed civil transport; significant advances are required to reduce takeoff gross weight and allow for both economic attractiveness and environmental acceptability."[56]

Since 2000 new industrial processes and materials have encouraged hopes that an SST, whether a small business plane or larger airliner, can now be made lighter and operate more economically. Novel approaches are being explored to reduce both the sonic boom and engine noise.[57] And recently a detailed market analysis forecast that ample demand for supersonic aircraft will emerge in the decades ahead.[58] Perhaps.

In the meantime proposals and fledgling development efforts are proliferating. Recently, for example, the Japanese Aerospace Exploration Agency (JAXA) has been developing an SST that would have only a mild sonic boom. The agency aims to build a small, lightweight plane that can carry 36–50 business travelers at Mach 1.6. The design concept features two engines and an elongated nose that comes to a sharp point. JAXA has an even more ambitious project in the works: a Mach 5 hypersonic aircraft whose prototype engine has already been undergoing tests.[59] One wonders about the fairness of a government-funded aircraft that would benefit only elite travelers.

Figure 12.3. NASA drawing of nexgen SST.
Source: courtesy of NASA.

And indefatigable NASA has also not given up. On February 29, 2016, the agency announced that work had begun "to build a quieter supersonic passenger jet."[60] The news release even included an artist's rendering of such an aircraft (Figure 12.3). A $20 million contract was awarded to Lockheed Martin—one of America's largest military contractors—for preliminary design work focused on converting the sonic boom to a mild sonic "bump" and, presumably, lessening engine noise on takeoffs and landings. NASA's faith that Lockheed Martin can confect a design that meets all modern performance requirements and promises commercial success may be little more than wishful thinking.

Despite this enthusiasm for a new generation of SSTs, no major aircraft manufacturer in the world has initiated a program to develop a supersonic airliner with its own money. In the meantime projects by JAXA, NASA, and others are furnishing invaluable experience for a new generation of aeronautical engineers that can sustain the dream of crossing the oceans faster than the speed of sound.

✳ ✳ ✳ ✳ ✳

Concorde aircraft and Tu-144s are on display in many museums.[61] In the eastern United States, one Concorde is at the Sea–Air–Space Museum in New York City and another at the Udvar–Hazy Center of the Smithsonian National Air and Space Museum, near Washington Dulles International Airport in Virginia. On the West Coast, the Concorde can be seen at the Museum of Flight in Tukwila, Washington, south of Seattle.

In England the Concorde is exhibited at the Bristol Filton Airport in Bristol and the Manchester Airport Viewing Park, and in Scotland at the National Museum of Flight in East Lothian.

France shows off the Concorde at the Airbus factory in Toulouse and at Orly Airport and the Le Bourget Air and Space Museum in Paris.

SST enthusiasts can see both the Concorde and the Tu-144 mounted on the roof of the Autoand Technik Museum in Sinsheim, Germany.

In Russia the Tupelov Tu-144 is exhibited at the Central Air Force Museum of Russia in Monino, the Museum of Civil Aviation in Ulyanovsk, and the Museum of Samara State Aerospace University.[62]

Boeing's mockup of the 2707-300 is being restored at the Museum of Flight near Seattle.[63]

NOTES

[1] Glancey (2015:47–48).

[2] https://en.wikipedia.org/wiki/De_Havilland_Comet, accessed 4 June 2016.

[3] Glancey (2015:9). Many books tell the Concorde story, but I rely mainly on Glancey.

4 Trubshaw (2000:12).

5 Glancey (2015:20).

6 Glancey (2015:22).

7 The complex political context of the SST is treated at length by Owen (1997).

8 Glancey (2015:43).

9 Glancey (2015:27).

10 Amery quoted in Glancey 2015:26.

11 Trubshaw (2000:40–47).

12 Glancey (2015:76).

13 Glancey (2015:59).

14 Glancey (2015:63). Dietrich Küchemann came up with this radical concept.

15 Trubshaw quoted in Glancey (2015:69).

16 On the HP.115 see also https://en.wikipedia.org/wiki/Handley_Page_HP.115, accessed 23 May 2016.

17 Glancey (2015:70–71); see also https://en.wikipedia.org/wiki/Fairey_Delta_2, accessed 23 May 2016.

18 Glancey (2015:70).

19 Glancey (2015:72–73).

20 Trubshaw 2000:51–52).

21 Glancey (2015:105).

22 Owen (1997).

23 Baldwin quoted in Glancey (2015:75).

24 Owen (1997:3).

25 Glancey (2015:89–90).

26 The $3 billion figure comes from Owen (1997:5).

27 Glancey 2015:66.

28 http://millercenter.org/president/kennedy/speeches/speech-5763, accessed 5 June 2016.

29 Owen (1997).

30 Witkin (1970).

31 Owen (1997:5).

32 For information on the Boeing 2707-300, I rely on Boeing (1969).

[33] Boeing (1969:1–1)

[34] Owen (1997).

[35] https://en.wikipedia.org/wiki/Boeing_2707, accessed 28 May 2016.

[36] Information on the Soviet SST comes mainly from https://en.wikipedia.org/wiki/Tupolev_Tu-144, accessed 5 June 2016.

[37] Shabad (1969).

[38] Shabad (1969).

[39] Glancey (2015:109).

[40] "Accident Seen as Major Blow to Soviet Hope for World Sales," *New York Times*, 4 June 1973, p. 19.

[41] Glancey (2015:112).

[42] "Concorde Tops Speed of Sound for 9 Minutes on a Test Flight," *New York Times*, 2 October 1969, p. 1.

[43] Trubshaw (2000:52).

[44] Glancey (2015:134).

[45] Hess (1971).

[46] Witkin (1973a).

[47] Quoted in Witkin (1973b).

[48] https://en.wikipedia.org/wiki/Concorde, accessed 6 June 2016.

[49] Glancey (2015:149).

[50] Glancey (2015:152).

[51] Collins (1978); "Concorde Still Losing Money," *New York Times*, 22 February 1980, p. A11.

[52] Trubshaw (2000:167).

[53] Trubshaw (2000:35).

[54] Glancey (2015:167).

[55] On cultural imperatives see Schiffer (2011:65–67).

[56] http://archive.org/stream/nasa_techdoc_19890018276/19890018276_djvu.txt, accessed 6 June 2016.

[57] Candel (2004) reviews the Concorde's technological innovations and outlines the daunting technical challenges that remain for a new generation of SSTs.

[58] Margaretic and Steelant (2015).

[59] http://www.aero.jaxa.jp/eng/research/frontier/sst/concept.html, accessed 9 June 2016.

[60] http://www.nasa.gov/press-release/nasa-begins-work-to-build-a-quieter-supersonic-passenger-jet, accessed 6 June 2016.

[61] Information on Concorde locations from http://airwaysnews.com/blog/2013/10/25/concordes-gone/, accessed 9 June 2016.

[62] https://en.wikipedia.org/wiki/Tupolev_Tu-144#Aircraft_on_display, accessed 9 June 2016.

[63] https://en.wikipedia.org/wiki/Museum_of_Flight, accessed 9 June 2016.

13

FUSION, HOT AND COLD

During World War II the United States government lured many distinguished physicists to Los Alamos, a remote region in northern New Mexico. As members of the Manhattan Project, they were tasked to help end the war by building the first atomic bombs. In an atomic bomb, uranium-235 or plutonium-239 fissions into lighter elements, generating radiation and neutrons; the latter then unleashes an uncontrolled chain reaction and enormous amounts of energy.

Edward Teller, a Los Alamos physicist brimming over with strong views, opined that fission bombs weren't sufficiently destructive. A fusion bomb, he insisted, would be much more powerful. Teller received permission to begin design work on a fusion bomb, nicknamed the "Super." Also called a hydrogen bomb (H-bomb) or thermonuclear device, a fusion bomb would create a huge burst of thermal energy when atoms of hydrogen fused to become helium—precisely the process that powers the sun.

The Manhattan Project did not build an H-bomb, although Teller, Stanislaw Ulam, and others eventually developed important design ideas. They knew that the biggest obstacle to earth-bound fusion is that atomic nuclei, which have a positive charge, strongly repel each other. To overcome this repelling force, hydrogen atoms must be greatly compressed and heated; this increases the intimacy between their nuclei, promoting fusion.[1] Creating these conditions requires a vast amount of energy, but Teller and Ulam had a bold idea for how to obtain it. They proposed that a fission bomb's energy could be concentrated on a small mass of hydrogen, whose nuclei would be forced to fuse. In this design a fission bomb sires the Super. But would it work?

The Super's design would soon be tested—the opportunity arising after the Soviet Union exploded its first atomic bomb in 1949, which led President Truman to start a crash program to build the Super. With Teller's involvement (he became known as "the father of the H-bomb") the program culminated on Eniwetok Atoll, a chain of coral islands in the Marshall group that the United States had taken from Japan.

Under the code name Operation Ivy, a government team assembled a Super on an Eniwetok isle after coercing its residents to leave. Twenty miles away on a U.S. Navy vessel, officers detonated the Super—nicknamed Ivy Mike—on November 1, 1952. The chilling video of the explosion shows an orange fireball three miles in diameter, which rapidly turned into a mushroom cloud one hundred miles wide and 108,000 feet high.[2] As for the island, it was left with a stunning amount of radioactivity and a deep crater more than a mile across.[3] The Super was estimated to have a yield equal to more than ten megatons of TNT, about five hundred times the power of the atom bomb dropped on Nagasaki and enough to obliterate all of Lower Manhattan or downtown Washington.

Could the awesome power of nuclear fusion be harnessed in a power reactor? Early on, confident that they would learn how to tame nuclear fusion for the benefit of humankind, physicists believed the answer to this question was yes. But the engineering challenges were, and are, immense.[4] It is necessary to create a machine on earth that can heat hydrogen atoms to the temperature of the sun's interior. These conditions would form a hydrogen plasma— a hot soup of separated protons and electrons—in which the energetic nuclei approach each other at high speed and fuse. Clearly, running a reactor that can reach 150 million degrees Celsius would itself require a huge amount of energy. The holy grail of fusion researchers is to make a reactor that exceeds the "breakeven point," producing more energy than it uses.

Physicists are a well-organized, articulate community of scientists whose prestige and influence grew greatly after World War II; after all, they had developed radar and the atomic bomb. Arguing that research on fusion would speed the day when a power reactor could run on an unlimited supply of inexpensive fuel, physicists in industrial countries convinced government agencies to support their research. And their efforts succeeded, as governments funded generation after generation of experimental machines, each one larger and more expensive than the last. One day, physicists continued to promise, a machine would surpass the breakeven point and pave the way for commercial reactors and clean, cheap electricity.

In the United States beginning in the early 1950s, dozens of fusion machines were built in many configurations by universities, corporations funded by federal grants and contracts, and national laboratories—including Los Alamos in New Mexico and Oak Ridge in Tennessee.[5] This potent constituency, spending large sums of federal money on experimental machines and employing thousands of highly trained people, created a fusion juggernaut whose momentum would carry it into the 21st century. Much as particle physics, the fusion enterprise became "big science on steroids" in the United States and other industrial countries.

So far the most promising approach to controlled fusion employs a tokamak, a large doughnut-shaped tube (a torus) ringed with magnets that—in some designs—compress and heat hydrogen isotopes inside. Tokamaks built during the 1980s added to engineering science, providing a fund of knowledge showing that controlled fusion was feasible. But no machine approached continuous operation, much less breakeven.

In March 1989, while fusion researchers were analyzing their data and preparing grant proposals for the next wave of must-have machines, they received shocking news: a team of chemists had created fusion at room temperature in a tabletop device of extraordinary simplicity. If those reports were true, "cold" fusion had the potential to revolutionize energy production throughout the world, perhaps making "hot" fusion obsolete.[6]

Stanley Pons was a chemist at the University of Utah who received his Ph.D. from the University of Southampton where he met Martin Fleischmann, one of the world's foremost electrochemists. The two scientists became close friends and for five years collaborated on a secret project at Utah. They were content to carry out experiments, in no hurry to publish their findings, until they heard that Steven E. Jones—a physicist at Brigham Young University (BYU) in Provo, Utah—was also working on cold fusion.

When Pons and Fleischmann learned about Jones's work, they immediately suspected treachery. Jones had been a reviewer of a Pons–Fleischmann grant proposal submitted to the U.S. Department of Energy (DOE). Pons rashly accused Jones of stealing their ideas but Jones denied it. The dispute had the potential to harm the reputations of both scientists and detract attention from their research. To smooth things over, the scientists and presidents of the two universities held a meeting on March 6.[7]

Such a high-level meeting was unusual, if not unprecedented, in academia. It took place because much more was at stake than the researchers' reputations. Since at least the late 17th century, when Newton and Leibniz both claimed to have invented calculus, researchers have tried to establish priority for a discovery or invention to enhance their reputations and prestige. But beginning in 1980, when Congress passed the Bayh–Dole Act, establishing priority suddenly had enormous financial implications for U.S. universities.[8] Before passage of this act, inventions made on a federally funded project could be patented, but the government owned the patent. As a result, universities and other organizations had no incentive to apply for patents—an expensive process. Then came Bayh–Dole, which allowed patent rights to remain with the researcher's organization.[9]

Seeing opportunities to cash in on the inventiveness of their faculty, research universities hired patent attorneys and set up offices that solicited invention disclosures, turning the most promising ones into patent applications. It was in this context that the controversy over cold fusion reached the highest level at both universities. After all, if cold fusion were real and showed promise for producing power commercially, then the patent-holding institution would stand to gain millions—if not billions—of dollars in licensing fees.

The university presidents and researchers agreed at the meeting that on March 24 both teams would submit articles announcing their discoveries to *Nature*. But the University of Utah administration, intent on publicly claiming priority for the invention, broke faith with BYU by calling a press conference for March 23, when Fleischmann and Pons were to report their results. This public disclosure would not affect Utah's patent application, which had been submitted earlier.[10] And Pons and Fleischmann had already sent a paper to the *Journal of Electroanalytical Chemistry*, which issued a suspiciously rapid acceptance that almost preceded the press conference. These moves not only infuriated Jones, but they also violated an important norm of physical science: physicists and chemists do not publicly announce their research results before a report has been formally accepted by a journal.

After being introduced to the packed hall by university officials, Pons and Fleischman gave brief statements followed by a question-and-answer period and then conducted a tour of their laboratory (Figure 13.1).[11] Fusion was said to be occurring in an electrolytic cell. In general terms an electrolytic cell consists of a glass container filled with a conductive liquid—the electrolyte. Two electrodes, often metal, are immersed in the electrolyte and connected to a source of direct current. When the power is passed through the electrolyte, positive ions are attracted to the negative electrode and negative ions to the positive one. Building an electrolytic cell is so simple that any reasonably handy person can make one at home, using a battery to generate hydrogen and oxygen from water (a weak electrolyte).

In the Pons–Fleischmann cell, the electrolyte was "heavy" water with a dash of a lithium salt. In a molecule of heavy water, the two hydrogen atoms—each consisting of a proton and an electron—are replaced by deuterium, a hydrogen isotope that also happens to have a neutron.[12] The electrodes were the rare metals platinum and palladium, the latter well known for its ability to absorb hydrogen. When the power was turned on, the heavy water decomposed. Oxygen was liberated at the platinum electrode and deuterium

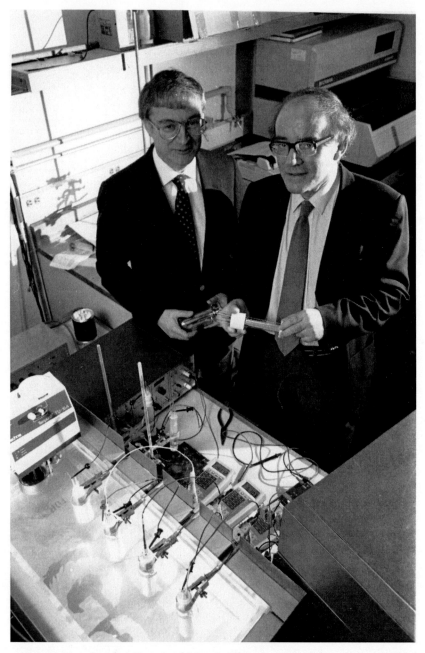

Figure 13.1. Stanley Pons and Martin Fleischmann show off their cold fusion apparatus. Source: courtesy of Special Collections, J. Willard Marriott Library, University of Utah.

SPECTACULAR FLOPS

atoms, attracted to the palladium, entered its crystal lattice. Pons and Fleischmann claimed that something special took place inside the palladium electrode: the deuterium atoms were severely compressed, enough to initiate nuclear fusion.

The main evidence for fusion was, as Pons told the audience, a "considerable release of energy . . . much more energy is coming out than we're putting in." This energy was in the form of heat and Pons said that the heat measurements "can only be accounted for by nuclear reactions . . . [not] by any chemical reaction that we know of." Fleischmann chimed in that the fusion reaction "can be sustained indefinitely. We have run experiments for hundreds of hours." Calculations from theory indicated that if nuclear reactions were occurring, then gamma rays, neutrons, some tritium (an even heavier isotope of hydrogen), and helium isotopes should also be produced. The chemists claimed to have observed gamma rays and tritium but, curiously, few neutrons and no helium. This was a puzzle, perhaps indicative of some sort of previously unknown fusion reaction—or experimental error.

During this unconventional report of research findings, Pons forecast that "it would be reasonable within a short number of years to build a fully operational device that could produce electric power—to drive a steam generator or steam turbine." While acknowledging that a richer base of science was needed before engineering issues could be tackled, Fleischmann did raise the possibility of creating power "with a relatively inexpensive device . . . fairly early on."

Although visitors to the Utah laboratory could see the experimental apparatus in plain view, other researchers would not be able to repeat the experiment faithfully until it was reported in detail. The *Journal of Electroanalytical Chemistry* published a preliminary note on April 10. In presenting data from four experiments that (purportedly) supported their press conference claims, Fleischmann and Pons emphasized that most of the excess heat was "due to an hitherto unknown nuclear process or processes."[13] Surprisingly, they projected that 1000 percent of breakeven was within reach.[14]

Scientists usually accept the published findings of their peers without actually repeating the experiments, but a revolutionary claim (such as cold fusion) demands replication. And replication requires a painstaking description of the original experiments so that they can be reproduced exactly; otherwise trials are apt to yield different or even conflicting results. The Fleischmann–Pons note was composed hastily and rushed into print under pressure from the university. The result was an incomplete account of what they had actually done.[15] Among the glaring absences were analyses of the composition and microstructure of the palladium electrode, as well as information about how much deuterium it had absorbed and the duration of that process. Also missing were graphs showing the cell's temperature over time and when, during the running of the experiments, the expected fusion products were measured. In short, despite the experiment's simplicity, researchers attempting to repeat it would have to guess at many crucial details.

The BYU researchers' report appeared in the April 27 issue of *Nature* and supplied more information than the Utah note. The eight authors took care to mention that they had begun this line of research in 1986, inspired by unknown geological processes that produced helium. To mimic natural conditions, their electrolyte contained—in addition to heavy water—a mixture of nine other compounds (metal salts) and a small amount of nitric acid. That fusion was occurring in their electrolytic cell was indicated by the production of neutrons, measured by a new instrument of their own design. In comparison with Fleishmann and Pons's exuberant claims, the BYU paper ends with a low-key statement. "Although the fusion rates observed so far are small, the discovery of cold nuclear fusion in condensed matter opens the possibility, at least, of a new path to fusion energy."[16]

After reading about cold fusion, hot fusion physicists scoffed and snickered. They knew that nuclear fusion, even in a tabletop device, would have produced enough radiation to quickly kill the researchers. But could Pons and Fleischmann have discovered a new kind of fusion that didn't produce as much radiation?

It didn't take long for other scientists (curious, skeptical, and hostile) to try replicating the cold fusion experiments—especially those of Pons and Fleischmann, which were simpler and had received much more publicity. Even before their note was published, experimenters around the world were gathering information about the Utah research from the press release, newspaper articles, and preprints on the Internet.[17] And the scientific press rushed into print with discussions of the cold fusion projects: *Nature* on March 30 published a very brief report, followed a day later by a longer one in *Science*. To prepare the latter article, Robert Pool interviewed some of the researchers and described the Pons–Fleischmann experiments.[18] From these varied sources, scientists around the world obtained enough information—imprecise though it was—to perform similar but not identical experiments. The result was a frenzy of activity in which scientific decorum was abandoned as researchers, following Pons and Fleischmann, first shared their results with the press.

The *New York Times* reported that "of the hundreds of groups around the country that have been struggling to repeat the Utah experiment, to date only Georgia Tech and Texas A & M University had claimed results that appeared" supportive of the Utah researchers.[19] In short order, however, both Georgia Tech and Texas A & M withdrew their confirmation claims.

In mid-April Pons and Fleischmann presented their research at the meeting of the American Chemical Society in Dallas, Texas, where thousands of members had gathered in a basketball arena to hear about cold fusion. In introducing the session, the society's president Clayton Callis pointed out that physicists were having major problems trying to transform hot fusion reactions into a practical energy source. Alluding to the cold fusion experiments, Callis went on to say that "chemists may have come to the rescue." The receptive audience laughed and clapped. According to the *Science* account, after hearing Pons and Fleischmann review their work, many in the audience were "willing to accept that it might be fusion."[20] That same article indicated that a few chemists and physicists—the most prominent being Peter Hagelstein of MIT—were working on theories to explain what was happening inside the palladium electrode, but there was no consensus in sight.[21]

During the first two days of May 1989, at the meeting of the American Physical Society in Baltimore, physicists took a turn assessing cold fusion. Fixed in their belief that cold fusion was impossible, physicists insisted that "assertions of 'cold fusion' were based on nothing more than experimental errors." Scientists from illustrious institutions—including MIT, Lawrence Berkeley Laboratory, Yale University, and Brookhaven National Laboratory—saw no "evidence of the existence of cold fusion." And it was said that "essentially all" experiments in Western Europe had failed.[22] At a news conference, a panel of nine researchers rendered a devastating judgment: cold fusion was dead.

More surprising still, Fleischmann and Pons were accused not only of doing error-laden experiments but also of incompetence, a cavalier charge that is rarely heard in scientific circles. And then the situation turned really nasty. The *Boston Herald* previewed the paper to be read by Ronald R. Parker, director of MIT's Plasma Fusion Center. He and his colleagues had concluded that the Utah experiments were "scientific schlock," "misrepresentation and 'maybe fraud.' "[23] This was just the beginning of the intense humiliation that Pons and Fleischmann would suffer at the hands of fellow scientists and the press.

When asked about the controversy, Pons told a reporter that "months of stinging criticism have taken their toll, leaving . . . [us] . . . frustrated and weary of meetings with skeptical and sometimes hostile colleagues. . . . They're saying it can't be done, but they can't explain our results."[24] Naming no names, he added that "many scientists who have achieved fusion are holding back their results for fear of public humiliation." Indeed, the term "cold fusion" was already heavily stigmatized. Surprisingly, one of the few physicists who took the cold fusion experiments seriously was Edward Teller, who called for more research.[25]

One of the most thorough attempts to repeat the Pons–Fleischmann experiments took place at the California Institute of Technology. A team of nineteen researchers, mainly chemists, published their findings in *Nature* on August 17, 1989.[26] Owing to uncertainties about how the Utah experiment was done, the team set up a dozen electrolytic cells, varying the palladium electrode and run times. They did observe excess heat but their highly sensitive instruments detected no helium, neutrons, tritium, or gamma rays above background levels. Instead of accounting for the excess heat by nuclear fusion, they invoked chemical processes. The report suggested that one source of error in the Pons–Fleischmann work was a failure to stir the heavy water while taking its temperature. The report also implied that the Utah researchers did not use state-of-the-art instruments to measure fusion products. Appearing in a prestigious journal, this report added to the large list of replications that failed to support the Utah claim for tabletop fusion.

Discounting the mounting criticism, the entrepreneurial University of Utah moved ahead quickly—hiring a Washington lobbying firm famed for using earmarks, a process that avoided normal reviews, to secure congressional appropriations. The lobbyists' pitch was for $25–40 million to help build a cold fusion research center in Utah. The university even hired Boston consultant Ira C. Magaziner to raise the alarm in Congress "that the Europeans, Japanese, and Koreans will steal America's latest invention, cold fusion, unless the federal government embarks on a crash program to understand the phenomenon and develop marketable technologies."[27] Congress held a hearing, but swayed by naysaying physicists provided no funding.[28]

Closer to home the university enjoyed more success. Utah Governor Norman H. Bangerter called a special session of the State Legislature and urged its members to fund a formal cold fusion program.[29] Acceding to the governor's request in mid-August 1989, the legislature founded the nonprofit National Cold Fusion Institute at the University of Utah with an appropriation of $4.5 million.[30] That the reality of cold fusion was in grave doubt did not deter the legislature from releasing the funds. By the end of October, the institute was up and running.

American scientists at other institutions could apply for cold fusion grants from the Electric Power Research Institute or the U.S. Department of Energy. The latter had granted Pons and Fleischmann more than $300,000 and the BYU team $1.9 million.[31]

Taking a special interest in cold fusion, DOE convened a panel of 23 physicists and chemists "to evaluate the results of the cold-fusion experiments conducted to date." The panel reviewed publications, visited laboratories, took part in the Workshop on Cold Fusion in Santa Fe, New Mexico, and held five public meetings. The final report was delivered to Secretary of Energy James D. Watkins on November 26, 1989.[32]

The DOE report concluded that "the present evidence for the discovery of a new nuclear process termed cold fusion is not persuasive." However, the report also noted that a small minority of laboratories had found support for "the Utah claims of excess heat production, usually for intermittent periods." Curiously, those same laboratories failed to "find commensurate quantities of fusion products, such as neutrons or tritium."[33] Noting the unexplained but intermittent heat, the panel called for further experiments to identify its cause. And experiments did continue.

A few years later, the DOE panel's co-chair John R. Huizenga published an extensive review of cold fusion—referring to it, in his book's subtitle, as "the scientific fiasco of the century."[34] A cottage industry of scientists and science journalists, in books and articles published since the early 1990s, have arrived at equally harsh judgments. In *Voodoo Science: The Road from Foolishness to Fraud*, physicist Robert L. Park used cold fusion as one of his case studies, noting that cold fusion "is dead, but the corpse won't stop twitching." He went on to argue that cold fusion was a phenomenon perpetuated by "inept scientists . . . greedy university administrators . . . gullible politicians . . . and careless journalists . . . all [with] an interest in pretending the issue had not been settled."[35] Park's explanation for why the corpse still twitched was self-interested actors, not new—and unexplained—science.

One of the more thoughtful postmortems was written by Denis Rousseau, a physical chemist. In *American Scientist* he maintained that cold fusion was a classic case of "pathological science." According to Rousseau:

> Pathological science arises from self-delusion—cases where scientists believe they are acting in a methodical, scientific manner but instead have lost their objectivity. The practitioners of pathological science believe that their findings simply cannot be wrong.[36]

Researchers who practice pathological science (according to Rousseau) are apt to reject accepted theories, attach great significance to barely detectable effects, fail to perform critical experiments, and find fault with any experiment that undermines the original claims.

When Rousseau published this article in 1992, his views reflected an overwhelming consensus. Most people in the press and in the scientific community believed that there was no definitive evidence for cold fusion—and probably never would be. In mainstream science the door of cold fusion, opened so brashly by Pons and Fleischmann, was by the end of the century slammed shut.

The National Cold Fusion Institute at the University of Utah, unable to secure outside funding after state money ran out, closed on June 31, 1991.[37] The following year Pons suddenly left the university and moved to France, where he and Fleischmann worked on cold fusion in a laboratory sponsored by Toyota. In 1998 Toyota, having spent £12 million without seeing conclusive results, closed the laboratory.[38]

But the corpse was still twitching vigorously. In fact, hundreds of researchers in the United States, India, Japan, Italy, and other countries continued to experiment—trying to improve reproducibility and explain the excess heat and traces of fusion products sometimes observed. Throughout the 1990s their work survived on the far fringes of mainstream science, their reports shunned by major journals. Even so, hundreds of papers on cold fusion were published in lesser journals and conference volumes during the decade after the phenomenon's alleged demise.

Early in the new century, researchers abandoned the toxic term "cold fusion" and substituted "low energy nuclear reactions" (LENR). A group of LENR researchers approached DOE in 2003 seeking a review of the evidence that had accumulated since the 1989 review. Agreeing to the request, DOE opted for an unusual review process.[39] DOE invited the researchers who had made the request to prepare a document that "identified the most significant experimental observations and publications." Peter L. Hagelstein and four other authors sent the report to DOE in 2004. In turn, DOE then distributed this document and supporting publications to nine scientists qualified to review it. In a second stage, DOE invited six LENR scientists to report on their research to another group of nine scientists.

Digesting the review of reviews, DOE's final report offered two major conclusions.[40] As for excess heat, "Most reviewers . . . stated that the effects are not repeatable, the magnitude of the effect has not increased in over a decade of work, and that many of the reported experiments were not well documented." Regarding actual evidence for nuclear reactions, "The preponderance of the reviewers' evaluations indicated that . . . the occurrence of low energy nuclear reactions, is not conclusively demonstrated." In short, nothing much had changed since the 1989 DOE report. The intermittent bursts of heat in the electrolytic cells remained unexplained and measurements of fusion products were still erratic and disputable. However, the report also recommended—as had its predecessor—that further research be undertaken.

Although Rousseau's concept of pathological science seems applicable to the more recent work in LENR, current researchers have offered some explanations for the inability of experiments to consistently produce excess heat. One explanation focuses on the palladium's chemical composition and on how it was prepared; another suggests that in failed experiments there was insufficient loading of the deuterium into the palladium.[41] And new explanations have been offered for why, in a palladium electrode, deuterium might fuse.[42]

Experiments in LENR are continuing and a few reports have been published in major journals. What's more, some LENR research has been funded by reputable organizations—including SRI International, a nonprofit research and consulting firm headquartered in Menlo Park, California, and the U.S. Naval Surface Warfare Center–Dahlgren Division (NAVSEA) in Virginia. More surprising still, in a 2015 Power Point presentation physicist Louis F. DeChiaro of NAVSEA listed ten entrepreneurs who are attempting to commercialize LENR phenomena.[43] Whether any of these purported technologies will materialize can only be shown by rigorous, independent evaluation.[44] For critics the odor of fraud hangs heavily in the air.

Cold fusion has risen from the dead, perhaps temporarily, but its rebound has not affected the pursuit of hot fusion. Experimental fusion reactors continue to be built, many of them novel designs, some funded by venture capitalists.[45] The most ambitious project is situated near Cadarache, in the south of France, where an international team of two thousand people is building the world's most advanced and largest tokamak (at 23,000 tons). The idea was conceived by Russia's leading fusion researcher Evgeny Velikhov, who had the ear of USSR General Secretary Mikhail Gorbachev. Gorbachev mentioned it to President Reagan at a 1985 meeting in Geneva and an agreement was reached to create "a collaborative international project to develop fusion energy for peaceful purposes."

In later meetings a multinational partnership was formed that allocated tasks and contributions. Costs are shared among 28 countries in the European Union along with China, India, Japan, Korea, Russia, Switzerland, and the United States. Together these 35 countries make up the organization known as the International Thermonuclear Experimental Reactor (ITER). [46]

As a megaproject with so many players, ITER took shape slowly, beginning with general concepts that gradually evolved into a detailed design in 2001. Beyond its sheer size, this tokamak has some interesting features. The toroidal vacuum chamber holding the plasma is surrounded by dozens of D-shaped, superconducting electromagnets that shape and contain the plasma (Figure 13.2). Each magnet is wound with niobium-tin or niobium-titanium wire and weighs 310 tons. At the center of the machine is a huge cylindrical electromagnet for heating the plasma (Figure 13.2). Neutrons created by the fusion process will bombard and heat the tokamak's stainless steel walls. The heat will be conducted to cooling towers and dissipated, but in a commercial reactor it would produce steam to drive turboelectric generators. The tokamak also requires a massive infrastructure to cool the superconducting magnets to minus 269 degrees Celsius, establish a near-perfect vacuum in the torus, remove heat from the torus's walls, and monitor all the reactor's processes (Figure 13.3).

The site was chosen in 2005 and construction began in 2010. Ten million parts and one million components are being developed and

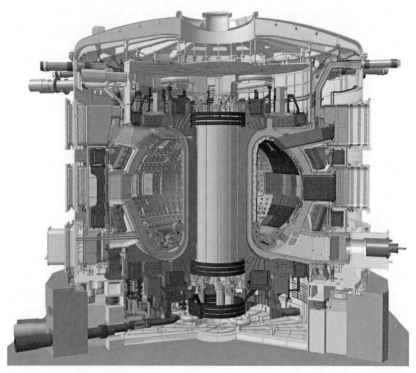

Figure 13.2. Cutaway drawing of the ITER tokamak.
Source: courtesy of ITER.

manufactured by contractors around the world. Completion of the facility is expected in 2025, but years of component and subsystem testing will be necessary before full-scale operation can begin in 2035. The goal is to beat breakeven by a factor of 10.

ITER's initial cost estimate was £5 billion, then it ballooned to £13 billion; as of mid-2016 it stood at about £20 billion, excluding operating and decommissioning expenses.[47] It will probably go higher owing to unexpected problems that may require design changes. Despite these immense cost overruns, the ITER constituency of physicists, engineers, and innumerable corporate contractors has been able to maintain the support of participating governments.

Because ITER is a research reactor that will produce only brief pulses of power, next-generation machines called DEMOs will address issues relating directly to commercialization. The goals of

SPECTACULAR FLOPS

Figure 13.3. Cutaway drawing of the ITER tokamak showing its infrastructure. Source: courtesy of ITER.

demonstration power plants will include operating continuously or near continuously, exceeding breakeven by a factor of 30 or 40, converting heat to electricity, generating tritium for its own fuel, and simplifying the design. Construction of DEMOs—probably in several countries—is expected to begin "in the 2030s and operation in the 2040s," timeframes likely to be extended.[48]

More than sixty years ago, Ivy Mike shattered the tranquility of Eniwetok Atoll just as U.S. physicists were beginning their first fusion experiments, hoping to create a new and bountiful source of energy.[49] After building many generations of fusion machines, most physicists have placed their hopes on the enormous and devilishly complex ITER tokamak, despite delays and cost overruns. In the end ITER is a grandiose experiment that might pave the way to generating clean, carbon-free energy. Or it might become a spectacular flop.

NOTES

1 Glasstone and Lovberg (1975:4).

2 https://archive.org/details/operation_ivy, accessed 19 August 2016.

3 https://en.wikipedia.org/wiki/Ivy_Mike, accessed 19 August 2016.

4 Glasstone and Lovberg (1975:5).

5 Bromberg (1982:265–266).

6 According to Schiffer, Hollenback, and Bell (2003:207), the term "cold fusion" was first used by Benjamin Franklin to describe what he believed was a heatless electrical process for melting wire.

7 Close (1991).

8 https://en.wikipedia.org/wiki/Bayh%E2%80%93Dole_Act, accessed 24 August 2016.

9 Technically, only individuals can receive a U.S. patent. That is why a university or company researcher is awarded the patent but, through prior contractual agreement, surrenders the rights to the organization.

10 Close (1991:100).

11 For a heavily edited video of this event, see https://www.youtube.com/watch?v=6CfHaeQo6oU, accessed 18 August 2016. I transcribed the Pons and Fleischmann quotes from this video. The written press release was accessed on 18 August at http://www.colorado.edu/physics/phys3000/phys3000_sp16/handouts/64_ColdFusion PressRelease.pdf.

12 Deuterium occurs naturally in seawater, about one in 6,500 hydrogen atoms (Glasstone and Lovberg 1975:2).

13 Fleischmann and Pons (1989:308).

14 Fleischmann and Pons (1989:305).

15 According to Pool (1990), experimental details were withheld to protect the university's patent rights.

16 Jones et al. (1989:740).

17 Preprints were mentioned on 6 April: "Cold Fusion Causes Frenzy but Lacks Confirmation," *Nature* 338:447.

18 "Cold Results from Utah," *Nature* 338:364; Pool (1989b).

19 Broad (1989a).

20 Quotes in this paragraph are from Pool (1989b).

21 Hagelstein is mentioned in "Scientific Look at Cold Fusion Inconclusive," *Nature* 338:605.

22 Quotes in this paragraph are from Browne (1989).

23 Quoted in Tate (1989).

24 "'Cold Fusion' Research Enjoys a Boom in Utah," *New York Times*, 1 August 1989, p. 11.

25 Pool (1989b).

26 Lewis et al. (1989).

27 Crawford (1989:522).

28 Park (2000:92–98) discusses the congressional hearing.

29 Pusey (1991).

30 Broad (1989b).

31 Pool (1989a, 1989b); "Cold Fusion Causes Frenzy but Lacks Confirmation," *Nature* 338:447.

32 U.S. Department of Energy (1989), quote on p. III.

33 U.S. Department of Energy (1989), quotes on p. 1.

34 Huizenga (1993).

35 Park (2000:97); see also Taubes (1993).

36 Rousseau (1992:54).

37 http://archiveswest.orbiscascade.org/ark:/80444/xv23135, accessed 20 August 2016.

38 https://en.wikipedia.org/wiki/Stanley_Pons, accessed 28 August 2016.

39 Quotes in this and the following paragraphs are from the unpaginated DOE report (U.S. Department of Energy 2004): http://newenergy times.com/v2/government/DOE2004/DOE-CF-Final-120104.pdf, accessed 8 July 2018.

40 U.S. Department of Energy (2004).

41 Miles and McKubre (2014).

42 Tsyganov (2012).

43 https://www.lenr-forum.com/forum/index.php/Attachment/386-IEEE-brief-DeChiaro-9-2015-pdf/, accessed 29 August 2016.

44 Featherstone (2012) recounts the difficulties of getting independent confirmation that Andrea Rossi's "energy catalyzer" (E-Cat) actually produces energy continuously.

45 Clery (2017); Martin (2015).

46 Information about the ITER project and all quotes come from pages on the ITER website—for example, https://www.iter.org/proj/iterhistory, accessed 25 August 2016.

47 "ITER Nuclear Fusion Project Faces New Delay, Cost Overrun." https://www.iter.org/faq, accessed 12 September 2018.

48 On DEMO see https://www.iter.org/sci/iterandbeyond, accessed 9 September 2016.

49 Bromberg (1982:265–266).

14

CONCLUSION

Some Limited Generalizations

All technologies have a life history (or life cycle). Some life histories are long, as in Lower and Middle Paleolithic hand axes made for more than a million years; others are short, as in the original DeLorean DMC-12 automobile made for just two years (1981–1983). Both technologies reached consumers, but the preceding chapters have shown that the life histories of spectacular flops often ended much earlier. Thus, to structure the following discussion, I employ a complete, idealized life history consisting of five processes: invention, development, manufacture, adoption, and use.[1] This structure enables the presentation of process-related generalizations. An overarching generalization amply supported by the case studies is that a technology's actual life history may end during any process.[2]

The forward motion of a technology's life history is facilitated by a promoter: one or more people, companies, or government agencies that envision the technology's game-changing potential and communicate this potential to others. In addition to spearheading the project, the promoter's major role is to obtain resources for development and manufacture. Acquired from private or public sponsors, resources in modern times mean money, which in turn can buy other resources such as labor, materials, equipment, and workspace. But lest we forget, at a project's outset some resources cannot be bought—as in new engineering science, specialized skills, government approvals, and time.[3]

Although usually an inventor or engineer, the technology's originator may also play the promoter role. For his visionary steamships, Brunel was both the engineer and promoter; in the latter role, he convinced companies to bankroll his projects. Tesla too was at times both inventor and promoter, tapping wealthy New Yorkers for support. Pons and Fleishmann invented cold fusion and vigorously promoted their controversial technology. A prosperous originator may even play the sponsor role. The Samuda brothers funded the first test of their atmospheric railway system at Wormwood Scrubs.

Promoters may change during lengthy projects, especially those sponsored by a government. Thus U.S. Air Force officials, at first Curtis LeMay and then Donald J. Keirn, as well as engineer Andrew Kalitinsky, made the case for the nuclear-powered bomber. In addition to Keirn, promoters in later years included the project's corporate contractors—especially General Electric and Pratt & Whitney—and politicians representing states and localities where those corporate activities took place. Sponsors may also change as promoters scramble to piece together resources. A case in point is the *Great Eastern*, which foundered during construction but a new company was organized to sell shares and the project resumed.

Invention begins with the originator's vision of a technology that might be. As the idea is given more definite form, the technology comes to be described in terms of anticipated performance characteristics. The latter are its most important or distinctive behavioral capabilities—ones that solve a perceived problem or that set it apart from existing, potentially competing technologies. The atmospheric railway was expected to be safer than locomotive-powered railways and less expensive to build and operate. The *Great Eastern*'s major performance characteristic was anticipated to be the ability to carry enough coal for a round-trip between Britain and

Australia. The nuclear bomber was projected to have a nearly unlimited range. And the Concorde was supposed to carry passengers safely at supersonic speeds.

To help materialize the imagined technology, its originator may make drawings and models for the promoter to use in pitching the invention to potential sponsors. In publicizing his design ideas for the first Dymaxion house, perhaps to attract funds for building a full-scale prototype, Buckminster Fuller crafted an intriguing model and displayed it in Chicago's Marshall Field department store. Although garnering attention in the print media, his model attracted no sponsor and the 1929 Dymaxion house failed to reach development. This case—and innumerable stillborn inventions pictured in patents and on the pages of *Popular Mechanics* and other magazines—indicates that many spectacular flops occur during the invention process.[4]

Some projects do not make it past invention because the promoter is ineffective. Championing his own inventions, Buckminster Fuller only occasionally obtained sponsors. If inventors are indifferent or ineffective promoters, potential sponsors may not learn or care about their game-changing inventions. Also, a project may fail if its resource needs are so great that no sponsor is willing to support it.

Promoters who promise more than their technologies can ever deliver may discourage potential sponsors. In the most extreme cases, an ambitious technology may flop if its proposed operation violates scientific principles. When presented with a questionable invention, a likely sponsor may ask scientific authorities to assess whether it is precluded by scientific constraints. During the mid-19th century, Joseph Henry—the Smithsonian Institution's first secretary and America's foremost physicist—played the role of scientific authority, debunking fraudulent and unfeasible inventions.[5] As we saw in Chapter 4, however, Henry also overplayed this role in the case of the early electric motor, invoking economic issues that he misrepresented as scientific constraints. Then and now scientific authority has been vested in the U.S. Patent and Trademark Office, which today employs thousands of scientists and engineers to judge the feasibility of inventions.

A project condemned on scientific grounds usually finds no sponsors unless the originator or promoter has the resources to move the project forward anyway. Defying a minority of scientific critics who insisted that a nuclear bomber would never work, the project passed from concept to development because it had avid promoters (the AEC and air force) that could secure federal funds. Likewise, an originator may push forward a questionable project that has modest resource needs. Almost immediately after the invention of cold fusion was publicly proclaimed, physicists insisted it was impossible. Nonetheless, inconsistent experimental findings stirred hopes that there might be something to cold fusion after all. Because cold fusion apparatus is relatively inexpensive and easy to assemble, experiments on "low energy nuclear reactions" continue to this day in dozens of laboratories around the world.

Development (also called research and development or R & D) consists of activities, such as experiments and the creation of prototypes, whose projected outcome is design specifications for a technology that in principle performs as expected and can be manufactured. Performance characteristics are refined during development and new ones of greater specificity are added.

An ordinary product usually undergoes a succession of prototypes, each one inching the technology closer to meeting its performance requirements and preparing the way to manufacture. By contrast, an ambitious, one-off project can only test individual components and subsystems. When these perform up to expectations, full-scale manufacture—sometimes in lengthy stages, as in the ITER tokamak—may follow.

The lack of sufficient resources appears to be the most common and immediate cause of failure during development. This may come about because the promoter underestimated the kinds and quantities of resources needed. Sometimes this is a deliberate strategy, as in de Lesseps' deceitful promotion of the Panama Canal. Competing for a project, U.S. military contractors have at times employed this strategy, expecting the government to cover the inevitable shortfall. However, flawed estimates of resource needs may also result from the very nature of the most ambitious and novel projects, which eventually require new materials, components, manufacturing processes, and engineering science. As such, there is genuine ignorance at the project's outset as to the entirety of specific resources needed to achieve success. One of the most difficult resources to estimate is time: how long will it take to complete development?

Technologies seemingly feasible on the basis of the simplest scientific principles have an allure that may be hard to resist despite overwhelming engineering challenges. The nuclear bomber was based on two simple principles: (a) nuclear reactors generate immense amounts of heat and (b) turbojet engines require immense amounts of heat. However, mating these two technologies in an aircraft required an inordinate amount of new resources. Even after more than a decade of development, a working nuclear bomber was still a fantasy. As long as U.S. government funds were forthcoming, the project lumbered along, sustained in part by the promoters' insistence that engineering problems—such as effective shielding for the crew—could be solved. But they were not solved before support ended.

I emphasize that running out of resources, especially funding, is merely the immediate cause of many spectacular flops. In fact, this dire outcome usually results from one or more contributing causes. The situation is much like that of the coroner who concludes that a person died from a massive heart attack but may also mention one or more contributing causes such as smoking, obesity, lack of exercise, and family history. And so it is with spectacular flops: we must ferret out the contributing causes in order to learn why resources were insufficient or unavailable to complete development.

In some cases development ends when the technology's function(s) is (are) no longer needed. Cugnot's steam *fardier* lost funding before the second prototype was thoroughly tested because its promoters had dropped out of the project and France, at peace, had no immediate need for a self-propelled dray. When U.S. defense requirements were reassessed in 1961, both the vast expense to continue work on the nuclear bomber (an estimated $1 billion) and the advent of intercontinental ballistic missiles led President Kennedy to kill the project. After all, the nuclear bomber lacked an important mission. Both the *fardier* and nuclear bomber suggest that a common contributing cause of failure during development (and manufacture) is obsolescence of the technology's intended function(s).

Another common contributing cause is an inability to achieve the technology's performance characteristics. A striking example is Charles Page's electric locomotive. After the federal grant ran out, he spent his own money, went into debt, and borrowed from friends. When at last he put the locomotive on its inaugural run, it performed poorly, a debacle that ended the project. Likewise, Brunel and Cubitt's long atmospheric railways—which included multiple pumping stations—were more expensive to build, operate, and maintain than locomotive-powered railways and had a limited hauling capacity. Accordingly, funding for these projects ended and no other long lines were built. When ongoing development reveals that the technology's performance requirements are unlikely to be met, the sponsor may be inclined to terminate the project.

An important contributing cause of many spectacular flops is an extended period of development during which a competing technology's performance characteristics are undergoing continuous improvement. Consequently, the new technology is aiming at a moving target it can't hit and may lose its sponsors. The Chrysler turbine engine competed against the internal combustion engine whose performance characteristics, especially gas mileage and tailpipe emissions, were improving dramatically. Turbine development continued for several decades and yielded many improved prototypes, but the turbine engine could not keep pace with the gasoline engine. Hopes for the atmospheric railway assumed that steam railways would remain slow and dangerous; instead they became faster, safer, and could easily add cars for hauling larger loads. It appears that a long-term project seeking to replace an existing technology is apt to fail if the latter's performance characteristics have greatly improved in the interim.

Some projects fail during development because they run headlong into technological constraints. It can be argued, for example, that even if Page had devised a highly efficient electric motor, his project would have failed anyway because batteries of the time were fragile and expensive to maintain. General Electric's "direct cycle" nuclear engine produced radioactive exhaust, a constraint that at best would have limited where the plane could take off and land.

Technological constraints depend on the state of contemporary engineering science. Thus well-funded projects with ample time can sometimes invent around the technical obstacles, creating new engineering science along the way. Huebner and his colleagues at Chrysler faced the constraint that a turbojet's exhaust is so hot that it would endanger anyone and anything near the tailpipe. They worked around this problem by inventing a "regenerator" that, through a heat exchanger, used exhaust gases to heat the air entering the combustion chamber. This solution lowered the temperature of the exhaust to a safe level and also improved the engine's efficiency. Clearly, technological constraints can be a showstopper or can incentivize a creative workaround.

Sometimes unforeseen scientific constraints do not derail a project until development is well along. The World System for transmitting power was based on faulty science stemming from Tesla's own inadequate experiments. No amount of tinkering with the Wardenclyffe transmitter would have allowed Tesla to broadcast power in the way he had envisioned. Ironically, Tesla's system might have sent information to long distances had he accepted the fact that he was generating Hertzian waves. But this would have meant conceding Marconi's contribution to wireless communication—anathema to Tesla. Unlike many scientifically unfeasible projects that find no sponsors early on, Tesla's World System survived, failing only during development.

Social constraints, as contributing causes, may also doom development (and manufacture). Project Chariot is a case in point. Teller and his Livermore group ignored the interests of Inupiat Eskimos who hunted and gathered around Ogotoruk Creek—land that was slated to be altered by nuclear blasts and harbor construction. Work to prepare the site at Ogotoruk Creek was far along when the project ran into the resolute opposition of Point Hope Eskimos, environmental scientists, and environmental organizations. In the face of this concerted opposition, the AEC put Project Chariot on hold and eventually abandoned it.

Social constraints also find expression in laws and regulations that, if imposed after development begins, can contribute to a project's demise. Federal regulations affecting tailpipe emissions and gas mileage helped cause the failure of the Chrysler turbine project.

✳ ✳ ✳ ✳ ✳

Manufacture (or commercialization) involves making the product so that it is available for adoption by consumers. As noted parenthetically above, several causes may lead to failures in development or manufacture.

In some cases the technology does not transition to manufacture for lack of a sponsor. Before initiating manufacture a company assesses the costs and benefits of bringing a new product to market—taking into account the estimated resource needs, performance characteristics of the technology and competing technologies, the company's resources and traditions, size of the potential market, expectations of profit, government regulations, and so on. If the benefits do not appear to justify the investment of resources, manufacture does not take place.

When Chrysler placed its fifty turbojet-powered cars in the hands of American consumers, hopes ran high that the stylish vehicle might soon be mass-produced. After driving the cars for three months, consumers had somewhat positive impressions. Despite this strong hint of potential demand, Chrysler chose not to mass-produce the car, perhaps swayed by several performance shortcomings: accelerator lag, poor gas mileage, and noxious exhaust emissions—the latter two coming under federal regulation. When the project began almost two decades earlier, automakers did not suspect they would have to meet gas mileage and emission standards. In this instance, and no doubt others, new social constraints affected the calculus of manufacturers.

Fuller's three prototype Dymaxion cars arrived in the early 1930s, an inauspicious time to put out a radical product. Despite a barrage of publicity, no automobile company signed on to make

Dymaxion cars. By this time automakers were addicted to incremental changes in annual models; oddball designs were generally eschewed because they did not seem to offer an expectation of profit. Fuller's car differed dramatically—in materials, appearance, drive-train layout, and so on—from anything else being produced at that time and didn't fit into any automakers' plans.

Sometimes promoters and manufacturers cannot agree on the terms to proceed. Through government funding Buckminster Fuller's second Dymaxion house, an update of his 1929 design, reached the prototype stage. In the postwar period, there was a severe housing shortage that Fuller's brainchild might have partially alleviated. Although several companies were interested in putting the house into production, Fuller refused permission, insisting that further development was needed—and it may have been so. Perhaps other quirky inventors afflicted with perfectionism also withheld their technologies from manufacture.

One-off projects that conduct development and manufacture simultaneously are especially risky and may be prone to fail. Among the case studies, de Lesseps' Panama Canal is a striking example. Insistent on building a sea-level canal, de Lesseps began construction before solving significant engineering problems such as taming the Chagres River and deepening the Culebras cut. The ITER tokamak may suffer the same fate.

Technologies rejected by manufacturers in one era may be commercialized decades later. The electric motor furnishes an interesting example. During the 1830s and 1840s, Thomas Davenport and Ransom Cook in the United States (as well as experimenters in Europe) developed powerful electric motors capable of replacing steam power to drive lathes, drills, printing presses, even small boats—yet no company made motors in quantity. Scientific authorities such as Joseph Henry disparaged electric motors because the batteries to power them were vastly more expensive to refurbish than refueling steam engines. These negative assessments, purportedly based on scientific constraints, seem to have discouraged potential manufacturers. In the late 1870s, when these (economic) constraints were relaxed by the commercialization of dynamos that could power motors inexpensively, many firms at last put electric motors into production and found a large market.

During *adoption* consumers have the opportunity to acquire the technology. In the case of a one-off technology commissioned by a company or government agency, adoption is highly probable as long as the project survives manufacture and achieves its expected performance requirements. However, when noncommissioned technologies become available to consumers, widespread adoption is far from assured.

Many technologies expire—sometimes quickly—after adoption begins. The Concorde's promoters were not trying to replace existing airliners but wanted to provide a unique service for well-heeled travelers. During the plane's long gestation period, however, Boeing and McDonnell Douglas both commercialized jumbo jets. These capacious aircraft were bought by most airlines and provided popular service, including first-class seats that cut deeply into the Concorde's customer base. Although dozens of airlines had placed orders for the Concorde early on, only the national airlines of France and Britain followed through with purchases. Airlines apparently judged that the likelihood of profit was slim in view of the dominance of jumbo jets and the Concorde's poor fuel economy and escalating price. Another contributing cause was local regulations on noise pollution that prevented the Concorde from landing at some important airports.

The Concorde case underscores the generalization that long-term development and manufacturing projects may be undercut by rapidly developing competitors and emerging social constraints. Both jumbo jets and environmental regulations hit the Concorde hard. Thus, despite being an impressive flying machine that performed its techno-function well, the Concorde became a spectacular flop.

* * * * *

Adoptions put the technology into consumers' hands and *use* follows almost immediately. A technology's performance during use determines how long manufacture and adoption continue. Having come this far, all technologies eventually fail and there are many contributing causes. Often changes in the social, cultural, economic, political, and environmental contexts render the technology's performance characteristics and functions (utilitarian, symbolic, or emotive) inadequate or obsolete, and adoption—and ultimately manufacture—come to an end. The cessation of manufacture is also known as a technology's senescence.[6]

Catastrophic failures and accidents during use may also hasten the end of manufacture and adoption. Poor reliability and several crashes led the Russians to discontinue making their supersonic airliner, the Tupolev Tu-144, and to permanently ground the fleet for passenger service.

The Great Ship Company's finances were overextended after the *Great Eastern*'s lengthy construction. Seeking a quick return on its investment, the company put the ship on the North Atlantic route although the vessel had been designed for travel to Australia. Dozens of capable steamships already provided regular Atlantic crossings and saturated the market; as a result, much of the *Great Eastern*'s vast capacity went unused. Suffering heavy losses, the company took the ship out of regular service as a passenger liner after less than three years.

An ambitious technology that succeeds showers fame on its originator and promoters, and prestige—in some cases earnings or profits—accrues to its sponsors. By contrast, a spectacular flop causes hardships and recriminations and may shame the people held responsible.

It appears that the more closely a project is identified with an individual, the greater its success or failure affects this person's reputation. Brunel built the Great Western Railway, an impressive achievement that put him in the ranks of Britain's foremost railway engineers. By virtue of his good reputation, charisma, and hubris, Brunel convinced the company to approve the atmospheric system for the South Devon line. Likewise, the success of several steamships gave Brunel credibility as a marine architect and he was able to secure the Eastern Steam Navigation Company's support for the *Great Eastern* project.

Let us generalize further: an inventor or engineer with a sterling reputation and persuasive skills may successfully promote a grand project that otherwise might not have been undertaken. This generalization seems unproblematic for the 19th century, the golden age of the independent inventor and heroic engineer. Yet I suggest that it also applies, although to a lesser extent, in recent times. Even in companies and government agencies, managers rely on the expertise and guidance of engineers who may promote a pet project and strongly influence development decisions. Would Chrysler have invested so much, for so long, in the turbojet project if Huebner— a highly respected company engineer—had not championed it? Would the AEC and Livermore have undertaken "geographical engineering" without Teller's prodding?

Spectacular flops can harm the reputation of promoters, sometimes even their health. In Brunel's later years, he promoted the South Devon atmospheric railway and the *Great Eastern*, acquiring a reputation for recklessly spending other people's money. The mental strains occasioned by the construction and launch of the *Great Eastern* likely hastened the great engineer's early death. The World System's failure led to Tesla's mental breakdown. And de Lesseps was left a broken man after his arrest and the failure of his Panama Canal company.

On a brighter note, spectacular flops often augment engineering science. A project may be envisioned on the basis of exquisitely simple scientific principles, but creating the necessary materials and manufacturing processes exposes gaps in knowledge. If the project is to move forward, these gaps have to be filled. This was clearly the case for the Chrysler turbine engine and nuclear bomber projects, both of which had to devise materials that could withstand high temperatures and mechanical strains. Experiments generated new principles of materials science that obviously informed the designs of turbojets and nuclear reactors, but these principles were also exploited by later projects. Likewise, development of the Concorde yielded new aeronautical principles. Even de Lesseps' Panama

Canal produced knowledge useful in the later American effort. The ongoing ITER project is providing a plethora of new engineering science for designing next-generation tokamaks. Although a spectacular flop may have been a nursery of new engineering science, a failed technology should not be rationalized on the basis of its spin-offs, which in all likelihood were neither predictable at the project's outset nor part of its original justification.

The most audacious technologies usually have symbolic functions that promote sociopolitical interests. As the largest ship in the world, the *Great Eastern* was a perfect symbol of Britain's domination of the high seas at a time when the sun never set on the British Empire. The Concorde, funded by the British and French governments, proclaimed that these two countries had—in the long shadow of World War II—mastered supersonic passenger flight, putting them at the forefront of civilian aviation technology. Project Chariot and other "atoms for peace" projects symbolized America's commitment to using nuclear technologies for nonlethal purposes. The nuclear bomber, on the other hand, sent the opposite message—that the air force, like the army and navy, was a competent player in the realm of nuclear weaponry. If we look hard enough, we may find that many (if not most) spectacular flops had important symbolic functions.

Projects having a long period of development, when resources may run short and technical problems may resist easy solutions, can be impelled forward if the sponsor deems their symbolic functions sufficiently important. Such anticipated functions may prolong development long after sober voices urge the project's termination. This was the case with the nuclear bomber, a political technology that encountered difficult—perhaps intractable—technical constraints such as radioactive exhaust. Even so, the project persisted, surviving the Eisenhower administration's skepticism. Despite massive cost overruns, the Concorde enjoyed continued support of the British and French governments because it was a symbol of their technological prowess. The Chrysler turbine project, which advertised the company's engineering excellence, continued to receive support despite the possibility that its performance shortcomings might never be remedied. And now that dozens of countries are invested in the ITER tokamak, a potent symbol of international collaboration, the project moves ahead even though nuclear fusion reactors might never produce a kilowatt-hour of clean, inexpensive electricity.

✳ ✳ ✳ ✳ ✳

Spectacular flops no doubt occurred in early Western and non-Western societies as long as there was a socioeconomic system capable of producing priests or kings or merchants with visions of ambitious technologies, who could command resources to launch their projects. That these prerequisites were met in many ancient societies is indicated by spectacular successes such as palaces, castles, temples, pyramids, huge fortifications, and large-scale irrigation systems. Alongside these highly visible successes, some projects also failed—as in the collapse of a new cathedral, an incomplete pyramid, and an unsuccessful canal. In addition, archaeologists and architectural historians sometimes uncover the remains of previous failures that were incorporated into later structures. Archaeologists are in an advantageous position to investigate the causes of early spectacular flops.

The case studies in this book reflect my interest in transportation, public works, and energy technologies. The question, of course, is whether the limited generalizations presented above apply to other kinds of technologies. I believe that the answer is yes because, as I maintain in other books, the processes of technological change apply across the board. And so I suggest that in other technological realms—including medical instruments, pharmaceuticals, scientific research equipment, agricultural tools, cooking utensils, home maintenance products, entertainment devices, and information—one will find spectacular flops that conform to this chapter's limited generalizations.

NOTES

1   On a technology's life history processes, see Hollenback and Schiffer (2010) and Schiffer (2011).

2   Assigning specific technological activities to a single process is sometimes difficult. In cases of ambiguous and coeval processes below, I employ the category that seems best for promoting analysis and understanding.

3   Schiffer (2011:86–88) furnishes a long list of resources a project might need.

4   Examples from *Popular Mechanics* in Benford (2010); see also Corn and Horrigan (1984).

5   Schiffer (2008a).

6   Hollenback and Schiffer (2010:324–325).

REFERENCES

Abbot, Henry L. 1907. *Problems of the Panama Canal.* Macmillan, New York.

Advisory Committee on Human Radiation Experiments. 1995. *Final Report of the Advisory Committee on Human Radiation Experiments.* U.S. Government Printing Office, Washington, DC.

Aitken, Hugh G. J. 1976. *Syntony and Spark: The Origins of Radio.* Princeton University Press, Princeton, NJ.

Albrecht, Ulrich. 1989. "The Nuclear-Propelled Bomber: A Faked Arms Race Between the US and USSR." In *Military Technology, Armaments Dynamics and Disarmament: ABC Weapons, Military Use of Nuclear Energy and of Outer Space and Implications for International Law,* edited by Hans Günter Brauch, pp. 127–164. St. Martin's Press, New York.

Atmore, Henry. 2004. "Railway Interests and the 'Rope of Air,' 1840–1848." *British Journal for the History of Science* 37:245–279.

Baldwin, J[ay]. 1996. *BuckyWorks: Buckminster Fuller's Ideas for Today.* Wiley, New York.

Banovsky, Michael. 2015. "Turbine Cars Are the Future That Never Arrived, Part 2." http://petrolicious.com/turbine-cars-are-the-future-that-never-arrived-part-2, accessed 2 April 2016.

Bauer, Reinhold. 2014. "Failed Innovations: Five Decades of Failure?" *Icon* 20:33–40.

Beatty, Charles. 1956. *De Lesseps of Suez: The Man and His Times.* Harper & Brothers, New York.

Beaver, Patrick. 1969. *The Big Ship: Brunel's Great Eastern – A Pictorial History.* Hugh Evelyn, London.

Beck, Colleen M., Susan R. Edwards, and Maureen L. King. 2011. *The Off-Site Plowshare and Vela Uniform Programs: Assessing Potential Environmental Liabilities Through an Examination of Proposed Nuclear Projects, High Explosive Experiments, and High Explosive Construction Activities.* Desert Research Institute, Cultural Resources Technical Report No. 111, Vols. 2–3.

Benford, Gregory. 2010. *The Wonderful Future That Never Was.* Hearst Books, New York.

Berry, Graham. 1957. "Nevadans Charge Fall-Out Danger; Livestock Fatalities Reported; Illnesses Blamed on Atom Tests." *Los Angeles Times*, 27 June, p. 1.

Bird, Kai, and Martin J. Sherwin. 2005. *American Prometheus: The Triumph and Tragedy of J. Robert Oppenheimer.* Knopf, New York.

Boeing Company. 1969. *The SST: A General Description.* Boeing Commercial Airplane Group, Supersonic Transport Division, Seattle, WA.

Bowles, Mark D. 2006. *Science in Flux: NASA's Nuclear Program at Plum Brook Station, 1955–2005.* National Aeronautics and Space Administration, Washington, DC.

Bramwell, Frederick. 1899. "The South Devon Atmospheric Railway, Preceded by Certain Remarks on the Transmission of Energy by a Partially Rarefied Atmosphere." Minutes of the Proceedings of the Meeting of the Institution of Mechanical Engineers, 25 July, pp. 299–327.

Braun, Hans-Joachim. 1992a. "Symposium on 'Failed Innovations.' " *Social Studies of Science* 22:213–230.

Braun, Hans-Joachim. 1992b. "The Chrysler Automotive Gas Turbine Engine, 1950–80." *Social Studies of Science* 22:339–351.

Broad, William J. 1989a. "Georgia Tech Team Reports Flaw in Critical Experiment on Fusion." *New York Times*, 14 April, p. D18.

Broad, William J. 1989b. "Despite Scorn, Team in Utah Still Seeks Cold-Fusion Clues." *New York Times*, 31 October, p. C1.

Bromberg, Joan L. 1982. *Fusion: Science, Politics, and the Invention of a New Energy Source*. MIT Press, Cambridge, MA.

Brooks, Paul, and Joseph Foote. 1962. "The Disturbing Story of Project Chariot." *Harper's Magazine*, 1 April, pp. 60–67.

Browne, Malcolm. 1989. "Physicists Debunk Claim of a New Kind of Fusion." *New York Times*, 3 May, p. A1.

Brunel, Isambard. [1870] 2006. *The Life of Isambard Kingdom Brunel, Civil Engineer*. Nonsuch, Gloucestershire, UK.

Buchanan, R. Angus. 2001. *Brunel: The Life and Times of Isambard Kingdom Brunel*. Hambledon Continuum, London.

Buck, Alice. 1983. "The Atomic Energy Commission." https://energy.gov/sites/prod/files/AEC%20History.pdf, accessed 22 December 2016.

Bush, Donald J. 1974. "Streamlining and American Industrial Design." *Leonardo* 7:309–317.

Bussard, R. W., and R. D. DeLauer. 1965. *Fundamentals of Nuclear Flight*. McGraw-Hill, New York.

Cain, Tubal. 1860. "The Great Eastern: First Voyage to the United States." *New York Times*, 29 June, p. 1.

Calon, Jean-Paul. 1994. "The Suez Canal Revisited, Ferdinand de Lesseps: The Genesis and Nurturing of Macroengineering Projects for the Next Century." *Interdisciplinary Science Reviews* 19:219–230.

Candel, Sébastien. 2004. "Concorde and the Future of Supersonic Transport." *Journal of Propulsion and Power* 20:59–68.

Carlson, W. Bernard. 2013. *Tesla: Inventor of the Electrical Age*. Princeton University Press, Princeton, NJ.

Cerf, Alain A. 2010. *Nicolas Cugnot and the Chariot of Fire*. Tampa Bay Automobile Museum, Pinellas Park, FL.

Cheney, Margaret. 1981. *Tesla: Man Out of Time*. Dorset Press, New York.

Chesebrough, Harry E. 1966. "Statement of Policy: Chrysler Corporation Turbine Car Program." Chrysler Corporation press release; on file, Transportation Collection, National Museum of American History, Smithsonian Institution, Washington, DC.

Chrysler Corporation. 1963. *The Chrysler Corporation Turbine Car*. Engineering Staff, Technical Information Section, Chrysler Corporation, Detroit, MI.

Chrysler Corporation. 1966. *History of the Chrysler Corporation Gas Turbine Vehicles, March 1954–June 1966*. Engineering Office, Technical Information, Chrysler Corporation, Detroit, MI.

Clancey, Jonathan. 2015. *Concorde: The Rise and Fall of the Supersonic Airliner*. Atlantic Books, London.

Claxton, Capt. (Chistopher). 1845. *History and Description of the Steam-Ship Great Britain Built at Bristol for the Great Western Steam-Ship Company*. J. Smith Homans, New York.

Clayton, Howard. 1966. *The Atmospheric Railways*. Self-published, Lichfield, UK.

Clegg, Samuel. 1839. *Clegg's Patent Atmospheric Railway*. Richard Kinder, London.

Clery, Daniel. 2017. "Private Fusion Machines Aim to Beat Massive Global Effort." *Science* 356:360–361.

Close, Frank. 1991. *Too Hot to Handle: The Race for Cold Fusion*. Princeton University Press, Princeton, NJ.

Collins, Joseph. 1978. "Concorde Loss for British Line." *New York Times*, 28 July, p. D1.

Colon, Raul. 2007. "Flying on Nuclear, the American Effort to Built [sic] a Nuclear Powered Bomber." The Aviation History Online Museum, 6 August. http://www.aviation-history.com/articles/nuke-american.htm, accessed 4 January 2016.

Conservatoire National des Arts et Métiers. 1956. *La Voiture à Vapeur de Cugnot, 1770*. Paris.

Constant, Edward, II. 1973. "A Model for Technological Change Applied to the Turbojet Revolution." *Technology and Culture* 14:553–572.

Corn, Joseph J., and Brian Horrigan. 1984. *Yesterday's Tomorrows: Past Visions of the American Future*. Summit Books, Washington, DC.

Cortright, Vincent. 1995. "Dream of Atomic Powered Flight." *Aviation History*, March, pp. 30–36, 69.

Cowper, Edward A. 1853. "Description of Cugnot's Original Invention of the Locomotive Steam-Engine for Common Roads." *London Journal of Arts, Sciences, and Manufactures* 42:301–303.

Crawford, Mark. 1989. "Utah Looks to Congress for Cold Fusion Cash." *Science* 244:522–523.

Cugnot, Nicolas-Joseph. 1769. *La Fortification de Campagne, Théorique et Pratique*. Paris.

Crocker, A. R. 1957. "Testing Aircraft Nuclear Power Plants." *Aeronautical Engineering Review* 16:30–35.

Culver, Henry B., and Gordon Grant. 1938. *Forty Famous Ships: Their Beginnings, Their Life Histories, Their Ultimate Fate*. Garden City Publishers, New York.

Davenport, Thomas. 1851. *Autobiography of Thomas Davenport*. Manuscript on file, Vermont Historical Society, Montpelier.

de Lesseps, Ferdinand. 1855. *The Isthmus of Suez Question*. Longman, Brown, Green & Longman, London.

de Lesseps, Ferdinand. 1876. *The History of the Suez Canal: A Personal Narrative*. An 1870 lecture translated by Henry Drummond Wolff. William Blackwood & Sons, Edinburgh, UK.

Dibner, Bern. 1959. *The Atlantic Cable*. Burndy Library, Norwalk, CT.

Emmerson, George S. 1977. *John Scott Russell: A Great Victorian Engineer and Naval Architect*. John Murray, London.

Emmerson, George S. 1981. *The Greatest Iron Ship: S.S. Great Eastern*. David & Charles, Newton Abbot, UK.

Engel, Leonard. 1958. "Twenty-Three Fishermen and a Bomb." *New York Times*, 23 February, p. BR1.

Farmer, Robert C. 1971. "The Automotive Gas Turbine: How Far Down the Road?" *Gas Turbine International* 11:14–27.

Featherstone, Steve. 2012. "Can Andrea Rossi's Infinite-Energy Black Box Power the World – Or Just Scam It?" *Popular Science*, 23 October, https://www.popsci.com/science/article/2012-10/andrea-rossis-black-box, accessed 30 June 2018.

Figuier, Louis. 1891. *Les Merveilles de la Science, Vol. 1*. Furne, Jouvet, et Cie, Paris.

Findlay, George. 1889. *The Working and Management of an English Railway*. Whittaker, Covent Garden, UK.

Fishbach, Laurence H. 1969. *Comparative Flight Envelopes for Three Different Design Point Subsonic Nuclear Airplanes.* NASA Technical Memorandum No. X–52619.

Fitzgerald, Percy. 1876. *The Great Suez Canal.* Savill, Edwards, London.

Fleischmann, Martin, and Stanley Pons. 1989. "Electrochemically Induced Nuclear Fusion of Deuterium." *Journal of Electroanalytical Chemistry and Interfacial Electrochemistry* 261:301–308.

Fletcher, Max E. 1958. "The Suez Canal and World Shipping, 1869–1914." *Journal of Economic History* 18:556–573.

Fletcher, William. 1891. *The History and Development of Steam Locomotion on Common Roads.* E. & F. N. Spon, London.

Flint, Jerry M. 1967. "Detroit Doubtful on Making Turbine Cars in the Near Future." *New York Times*, 4 June, p. 84.

Frenkel, Karen A. 2008. "Resuscitating the Atomic Airplane: Flying on a Wing and an Isotope." *Scientific American,* 5 December, http://www.scientificamerican.com/article.cfm?id=nuclear-powered-aircraft, accessed 30 June 2018.

Friedel, Robert. 2007. *A Culture of Improvement: Technology and the Western Millennium.* MIT Press, Cambridge, MA.

Fuller, R. Buckminster. 1932. "The Dymaxion House." *Fortune,* July, pp. 64–65.

Fuller, R. Buckminster. 1983. *Inventions: The Patented Works of R. Buckminster Fuller.* St. Martin's Press, New York.

Galloway, Elijah. 1826. *History of the Steam Engine: From Its First Invention to the Present Time.* Cowie, London.

Galloway, Robert Lindsay. 1881. *The Steam Engine and Its Inventors: A Historical Sketch.* Macmillan, London.

Gantz, Kenneth F. (editor). 1960. *Nuclear Flight: The United States Air Force Programs for Atomic Jets, Missiles, and Rockets.* Duell, Sloan & Pearce, New York.

Garthoff, Raymond L. 2016. "The *Swallow* and *Caspian Sea Monster* vs. the *Princess* and the *Camel*: The Cold War Contest for Nuclear-Powered Aircraft." *Studies in Intelligence* 60:1–11.

General Accounting Office. 1963. "Review of Manned Aircraft Nuclear Propulsion Program, Atomic Energy Commission and

Department of Defense." http://fas.org/nuke/space/anp-gao 1963.pdf, accessed 29 January 2016.

General Electric. 1962. "APEX-901: Comprehensive Technical Report, General Electric Direct-Air-Cycle Aircraft Nuclear Propulsion Program – Program Summary and References." General Electric, Flight Propulsion Laboratory Department, Cincinnati, OH. https://www.osti.gov/biblio/1048124, accessed 31 January 2016.

Glancey, Jonathan. 2015. *Concorde: The Rise and Fall of the Supersonic Airliner*. Atlantic Books, London.

Glasstone, Samuel, and Ralph H. Lovberg. 1975. *Controlled Thermonuclear Reactions: An Introduction to Theory and Experiment*. Krieger, Huntington, NY.

Gooday, Graeme. 1998. "Re-writing the 'Book of Blots': Critical Reflections on Histories of Technological 'Failure.' " *History and Technology* 14:265–291.

Graham, Frederick. 1947. "U.S. at Work on Project to Apply Atomic Power to Planes, Missiles." *New York Times*, 23 February, p. 43.

Gregory, R. H. 1982. *The South Devon Railway*. Oakwood Press, Salisbury, UK.

Hadfield, Charles. 1967. *Atmospheric Railways: A Victorian Venture in Silent Speed*. David & Charles, Newton Abbot, UK.

Haitch, Richard. 1966. "Today Just Never Arrives for the Car of Tomorrow." *New York Times*, 10 April, p. 368.

Harmon, Robert A. 1982. "Gas Turbines: Automotive Applications – The Outlook Behind the Wheel." *Mechanical Engineering* 104:26–45.

Hatch, Alden. 1974. *Buckminster Fuller: At Home in the Universe*. Crown, New York.

Hearn, Chester G. 2004. *Circuits in the Sea: The Men, the Ships, and the Atlantic Cable*. Praeger, Westport, CT.

Heck, J. G. 1851. *Iconographic Encyclopaedia, Division X: Technology*. Rudolph Garrigue, New York.

Henry, Joseph. 1831. "On a Reciprocating Motion Produced by Magnetic Attraction and Repulsion." *American Journal of Science and Arts*, pp. 340–343.

Hess, John L. 1971. "Concorde Jet Priced at Record $31.2-Million." *New York Times*, 16 December, p. 93.

Hewlett, Richard G., and Jack M. Holl. 1989. *Atoms for Peace and War, 1953–1961: Eisenhower and the Atomic Energy Commission*. University of California Press, Berkeley.

Hollenback, Kacy L., and Michael Brian Schiffer. 2010. "Technology and Material Life." In *The Oxford Handbook of Material Culture Studies*, edited by Dan Hicks and Mary Beaudry, pp. 313–332. Oxford University Press, Oxford, UK.

Howe, Henry. 1844. *Memoirs of the Most Eminent American Mechanics: Also, Lives of Distinguished European Mechanics*. Alexander V. Blake, New York.

Huebner, George J., Jr. 1964/1965. "The Chrysler Gas Turbine Story." *Institution of Mechanical Engineers – Automobile Division, Proceedings 179, Part 2A*, pp. 257–279.

Huebner, George J., Jr. 1966. "50-Car Turbine Program: Technical Results." Chrysler Corporation press release; on file, Transportation Collection, National Museum of American History, Smithsonian Institution.

Huebner, George J., Jr. 1979. *Energy and Pollution vs. Alternative Piston and Gas Turbine Powerplants*. Society of Automotive Engineers, Technical Series No. 790020.

Hughes, Thomas P. 1983. *Networks of Power: Electrification in Western Society, 1880–1930*. Johns Hopkins University Press, Baltimore, MD.

Huizenga, John R. 1993. *Cold Fusion: The Scientific Fiasco of the Century*. Oxford University Press, Oxford, UK.

Hunt, Robert. 1851. *Elementary Physics: An Introduction to the Study of Natural Philosophy*. Reeve & Benham, London.

Hunt, Robert. 1860. "Electro-Motive Engines." In *Ure's Dictionary of Arts, Manufactures, and Mines* (5th edition), edited by Robert Hunt, p. 100. Longman, Green, Longman & Roberts, London.

Israel, Paul. 1998. *Edison: A Life of Invention*. Wiley, New York.

Jacomy, Bruno, and Annie-Claude Martin. 1992. *Le Chariot à Feu de M. Cugnot*. Musée National des Techniques, Paris.

Johnson, C. L., and F. A. Cleveland. 1957. "Design of Air Frames for Nuclear Power." *Aeronautical Engineering Review* 16:48–57.

Jones, S. E., et al. 1989. "Observation of Cold Nuclear Fusion in Condensed Matter." *Nature* 338:737–740.

Joseph, James. 1961. "Details on the NX2: Our Atomic Plane." *Science and Mechanics*, January, pp. 66–70.

Kaiser, Walter. 2003. "Clean Air Act and American Automobile Industry." *Icon* 9:31–43.

Kalitinsky, Andrew. 1948. "Atomic Engines for Aircraft." *Pegasus* 16:1–4, 16.

Katz, Nathan. 1989. "Opportunities and Prospects for the Application of Structural Ceramics." In *Structural Ceramics*, edited by John B. Wachtman, Jr. *Treatise on Materials Science and Technology* 29:1–27.

Kaufman, Scott. 2013. *Project Plowshare: The Peaceful Use of Nuclear Explosives in Cold War America*. Cornell University Press, Ithaca, NY.

Keirn, Donald J. 1960. "The USAF Nuclear Propulsion Programs." In *Nuclear Flight: The United States Air Force Programs for Atomic Jets, Missiles, and Rockets*, edited by Kenneth F. Gantz, pp. 13–19. Duell, Sloan & Pearce, New York.

Kelly, Derek A. 1982. "The Philosophy of R. Buckminster Fuller." *International Philosophical Quarterly* 22:295–314.

Kenton, John. 1956. "Who's Who in the Aircraft Nuclear Program." *Nucleonics* 14:74–76.

Kirsch, David A. 2000. *The Electric Vehicle and the Burden of History*. Rutgers University Press, New Brunswick, NJ.

Kirsch, Scott. 2005. *Proving Grounds: Project Plowshare and the Unrealized Dream of Nuclear Earthmoving*. Rutgers University Press, New Brunswick, NJ.

Kline, Ronald. 1987. "Science and Engineering Theory in the Invention and Development of the Induction Motor, 1880–1900." *Technology and Culture* 28:283–313.

Klingensmith, Kenneth K., and Carl D. Lingenfelter. 1960. "Indirect-Cycle Nuclear Propulsion." In *Nuclear Flight: The United States Air Force Programs for Atomic Jets, Missiles, and Rockets*, edited by Kenneth F. Gantz, pp. 102–112. Duell, Sloan & Pearce, New York.

Kludas, Arnold. 2002. *Record Breakers of the North Atlantic: Blue Riband Liners 1838–1952*. Brassey's, Washington, DC.

Kopp, Carolyn. 1979. "The Origins of the American Scientific Debate Over Fallout Hazards." *Social Studies of Science* 9:403–422.

Krebs, Albin. 1983. "R. Buckminster Fuller Dead: Futurist Built Geodesic Dome." *New York Times*, 2 July, p. 1.

Lambright, W. Henry. 1967. *Shooting Down the Nuclear Airplane*. Bobbs-Merrill, Indianapolis, IN.

Langworth, Richard M., and Jan P. Norbye. 1985. *The Complete History of Chrysler Corporation 1924–1985*. Beekman House, New York.

Lehto, Steve. 2010. *Chrysler's Turbine Car: The Rise and Fall of Detroit's Coolest Creation*. Chicago Review Press, Chicago, IL.

Leslie, Stuart W. 1993. *The Cold War and American Science: The Military–Industrial–Academic Complex at MIT and Stanford*. Columbia University Press, New York.

Leupold, Jacob. 1724. *Theatri Machinarum Hydraulicarum, Tomus II*. Leipzig, Germany.

Lewis, Craig. 1958. "Nuclear Test Bomber Provided Valuable Shielding Data." *Aviation Week*, 22 December, pp. 64–69.

Lewis, E. B. 1957. "Leukemia and Ionizing Radiation." *Science* 125:965–972.

Lewis, N. S., et al. 1989. "Searches for Low-Temperature Nuclear Fusion of Deuterium in Palladium." *Nature* 340:525–530.

Lindsay, Forbes. 1912. *Panama: The Isthmus and the Canal*. John C. Winston, Philadelphia.

Ludvigsen, K. 1961. "Chrysler's Turbine for Today." *Car and Driver*, June, pp. 82–84, 87–89.

MacLaren, Malcolm. 1943. *The Rise of the Electrical Industry During the Nineteenth Century*. Princeton University Press, Princeton, NJ.

Magraw, Katherine. 1988. "Teller and the 'Clean Bomb' Episode." *Bulletin of the Atomic Scientists* 44:32–37.

Margaretic, Paula, and Johan Steelant. 2015. "Economical Assessment of Commercial High-Speed Transport." https://www.researchgate.net/profile/Johan_Steelant/publication/281641342_Economical_Assessment_of_Commercial_High-Speed_Transport/links/55f1f71908aef559dc49316e.pdf, accessed 11 June 2016.

Marks, Robert, and R. Buckminster Fuller. 1973. *The Dymaxion World of Buckminster Fuller*. Anchor/Doubleday, Garden City, NY.

Martin, Richard. 2015. "Finally, Fusion Takes Small Steps Toward Reality." *MIT Technology Review*, 14 September, https://www.technologyreview.com/s/541286/finally-fusion-takes-small-steps-toward-reality/, accessed 30 June 2018.

Maurer, Noel, and Carlos Yu. 2011. *The Big Ditch: How America Took, Built, Ran, and Ultimately Gave Away the Panama Canal*. Princeton University Press, Princeton, NJ.

McCullough, David. 1977. *The Path Between the Seas: The Creation of the Panama Canal 1870–1914*. Simon & Schuster, New York.

McLean, A. F. 1970. "Why Gas Turbines of the Future Will Run on Ceramics." *The Engineer* 231:29–31.

Mendizábal, Tomás. 2011. *Informe de Inspección Arqueológica de Activitades de Salvamento de Artefactos Arqueológicos Bordada Bohío, Lago Gatún*. Technical Evaluation No. 12, presented to the Canal Authority of Panama.

Miles, Melvin H., and Michael C. H. McKubre. 2014. "Cold Fusion After a Quarter-Century: The Pd/D System." In *Developments in Electrochemistry: Science Inspired by Martin Fleischmann*, edited by Derek Pletcher, Zhong-Qun Tian, and David Williams, pp. 245–260. Wiley, Chichester, UK.

Miller, David F. 1966. "Users' Evaluation of the Turbine Car." Chrysler Corporation press release; on file, Transportation Collection, National Museum of American History, Smithsonian Institution.

Miller, M. M. 1970. "Nuclear Airplane Now!" *Industrial Research* 12:48–50.

Molotsky, Irvin. 1994. "Washington-to-London Concorde Flights End." *New York Times*, 4 December, p. 3.

Mom, Giis. 2004. *The Electric Vehicle: Technology and Expectations in the Automobile Age*. Johns Hopkins University Press, Baltimore, MD.

Morin, M. 1851. "Note sur la Machine Locomotive de Cugnot, Déposée au Conservatoire des Arts et Métiers." *Comptes Rendus Hebdomadaires des Séances de l'Académie des Sciences* 32:524–532.

New Panama Canal Company. 1898. Untitled brochure in the library of the National Museum of American History, Smithsonian Institution.

Nicholls, James F., and John Taylor. 1882. *Bristol: Past and Present, Vol. 3*. Arrowsmith, Bristol, UK.

Norbye, Jan P. 1975. *The Gas Turbine Engine: Design, Development, Applications*. Chilton, Radnor, PA.

Norbye, Jan P. 1981. "Mercedes Gas Turbine: Is 16.5 mpg in a 3,500-lb. Car Good Enough?" *Popular Science*, March, pp. 23–24.

O'Neill, Dan. 2007. *The Firecracker Boys: H-Bombs, Inupiat Eskimos, and the Roots of the Environmental Movement*. Basic Books, New York.

O'Neill, John J. 1944. *Prodigal Genius: The Life of Nikola Tesla*. Ives Washburn, New York.

Otis, Fessenden N. 1862. *Illustrated History of the Panama Railroad*. Harper & Brothers, New York.

Owen, Kenneth. 1997. *Concorde and the Americans: International Politics of the Supersonic Transport*. Smithsonian Institution Press, Washington, DC.

Page, Charles G. 1845. "New Electro-Magnetic Engine." *American Journal of Science and Arts* 49:131–135.

Page, Charles G. 1867. *History of Induction: The American Claim to the Induction Coil and Its Electrostatic Developments*. Polkinhorn & Son, Washington, DC.

Page, Charles G. [1850] 1973a. "Report to William A. Graham, Secretary of the Navy, 30 August 1850." In *The New State Pa-*

pers: *Science and Technology, Vol. 13: Special Studies,* edited by Thomas C. Cockran, pp. 409–426. Scholarly Resources, Wilmington, DE.

Page, Charles G. [1851] 1973b. "Report to William A. Graham, Secretary of the Navy, 28 November 1851." In *The New State Papers: Science and Technology, Vol. 13: Special Studies,* edited by Thomas C. Cockran, pp. 64–67. Scholarly Resources, Wilmington, DE.

Park, Robert L. 2000. *Voodoo Science: The Road from Foolishness to Fraud.* Oxford University Press, Oxford, UK.

Pauling, Linus. 1958. "Fact and Fable of Fallout." *The Nation,* 14 June, pp. 537–542.

Pawley, Martin. 1990. *Buckminster Fuller.* Trefoil Publications, London.

Pawley, Martin. 2002. "The Downfall of the Dymaxion Car." In *Autopia: Cars and Culture,* edited by Peter Wollen and Joe Kerr, pp. 371–378. Reaktion Books, London.

Pool, Robert. 1989a. "Fusion Breakthrough?" *Science* 243:1661–1662.

Pool, Robert. 1989b."Teller, Chu 'Boost' Cold Fusion." *Science* 246:449.

Pool, Robert. 1990. "Only the Grin Remains." *Science* 250:754–755.

Post, Robert C. 1976. *Physics, Patents, and Politics: A Biography of Charles Grafton Page.* Science History Publications, New York.

Prokesh, Steven. 1990 "Study Set on Future of the Concorde." *New York Times,* 10 May, p. 6.

Pusey, Roger. 1991. "Bangerter Urging U. to Push Cold Fusion Within the School's Financial Capabilities." *Deseret News,* 25 May, https://www.technologyreview.com/s/541286/finally-fusion-takes-small-steps-toward-reality, accessed 30 June 2018.

Redfield, W. C. 1833. "Notices of American Steamboats." *American Journal of Science and Arts* 23:311–318.

Reeve, Arthur B. 1911. "Tesla and His Wireless Age." *Popular Electricity,* 1 June, p. 97.

Reines, Frederick. 1950. "Are There Peaceful Engineering Uses of Atomic Explosives?" *Bulletin of the Atomic Scientists* 6:171–172.

Ritchie, Robert. 1846. *Railways: Their Rise, Progress, and Construction – With Remarks on Railway Accidents, and Proposals for their Prevention*. Longman, London.

Rodrigues, J. C. 1885. *The Panama Canal: Its History, Its Political Aspects, and Financial Difficulties*. Sampson Low, Marston, Searle & Rivington, London.

Rom, Frank E., and Charles C. Masser. 1971. *Large Nuclear-Powered Subsonic Aircraft for Transoceanic Commerce*. NASA Technical Memorandum No. X–2386.

Rousseau, Denis L. 1992. "Case Studies in Pathological Science." *American Scientist* 80:54–63.

Roy, Amedee, Frederick A. Hagen, and Claude Belleau. 1964. "Chrysler's Gas Turbine Car: Materials Requirements." Paper presented at the Society of Automotive Engineers, Automotive Engineering Congress and Exposition, Detroit, Michigan, No. 777C.

Russell, J. Scott. 1864/1865. *The Modern System of Naval Architecture*. Day & Son, London.

Rybczynski, Witold. 1992. "A Little House on the Prairie Goes to a Museum: ARCHITECTURE VIEW . . ." *New York Times*, 19 April, p. 34.

Samuda, Joseph D'Aguilar. 1841. *A Treatise on the Adaptation of Atmospheric Pressure to the Purposes of Locomotion on Railways*. John Weale, London.

Savery, Thomas. 1699. "An Account of Mr. Tho Savery's Engine for Raising Water by the Help of Fire." *Philosophical Transactions of the Royal Society* 21:228.

Savery, Thomas. [1702] 1829. *The Miner's Friend: Or, an Engine to Raise Water by Fire, Described*. J. McCormick, London.

Schiffer, Michael Brian. 2000. "Indigenous Theories, Scientific Theories and Product Histories." In *Matter, Materiality and Modern Culture*, edited by P. M. Grave-Brown, pp. 72–96. Routledge, London.

Schiffer, Michael Brian. 2008a. *Power Struggles: Scientific Authority and the Creation of Practical Electricity Before Edison*. MIT Press, Cambridge, MA.

Schiffer, Michael Brian. 2008b. "Expanding Ethnoarchaeology: Historical Evidence and Model-Building in the Study of Technological Change." In *Oxford Handbook of Engineering and Technology in the Classical World*, edited by John Peter Oleson, pp. 821–835. Oxford University Press, Oxford, UK.

Schiffer, Michael Brian. 2010. *Behavioral Archaeology: Principles and Practice*. Equinox, London.

Schiffer, Michael Brian. 2011. *Studying Technological Change: A Behavioral Approach*. University of Utah Press, Salt Lake City.

Schiffer, Michael Brian. 2013a. *The Archaeology of Science: Studying the Creation of Useful Knowledge*. Springer, New York.

Schiffer, Michael Brian. 2013b. "Afterlives." In *The Archaeology of the Contemporary World*, edited by Paul Graves-Brown, Rodney Harrison, and Angela Piccini, pp. 247–260. Oxford University Press, Oxford, UK.

Schiffer, Michael B., Tamara C. Butts, and Kimberly Grimm. 1994. *Taking Charge: The Electric Automobile in America*. Smithsonian Institution Press, Washington, DC.

Schiffer, Michael B., Kacy L. Hollenback, and Carrie Bell. 2003. *Draw the Lightning Down: Benjamin Franklin and Electrical Technology in the Age of Enlightenment*. University of California Press, Berkeley.

Schiffer, Michael Brian, Charles Riggs, and J. Jefferson Reid (editors). 2017. *The Strong Case Approach in Behavioral Archaeology*. Univerity of Utah Press, Salt Lake City.

Schuwer, T. J. H. S. 2015. *Feasibility Study of a Nuclear-Powered Passenger Aircraft: Heat Cycle Design for the RECREATE Cruiser*. M.S. Thesis, Aerospace Engineering, Delft University of Technology.

Seaborg, Glenn T., and William R. Corliss. 1971. *Man and Atom: Building a New World through Nuclear Technology*. E. P. Dutton, New York.

Select Committee. 1845. *Report from the Select Committee [of the House of Commons] on Atmospheric Railways: Together with the Minutes of Evidence*. House of Commons, London.

Seren, Leo. 1958a. "What Are the Safety Hazards of Nuclear Propulsion?" *Space Aeronautics* 30:18–19.

Seren, Leo. 1958b. "Safety Hazards of Nuclear Propulsion – II." *Space Aeronautics* 30:48–52.

Shabad, Theodore. 1969. "Russian Supersonic Airliner Gets Test Flight, Beating the Concord." *New York Times*, 1 January, p. 1.

Skibo, James M., and Michael Brian Schiffer. 2008. *People and Things: A Behavioral Approach to Material Culture.* Springer, New York.

Skibo, James M., William H. Walker, and Axel Nielsen (editors). 1995. *Expanding Archaeology.* University of Utah Press, Salt Lake City.

Smiles, Samuel. 1868. *The Life of George Stephenson and of His Son Robert Stephenson: Comprising also a History of the Invention and Introduction of the Railway Locomotive.* Harper & Brothers, New York.

Smith, Frederick, and Peter Barlow. 1842. *Report to the Right Honourable the Earl of Ripon, President of the Board of Trade, on the Atmospheric Railway.* Her Majesty's Stationery Office, London.

Snyder, Bernard J. 1996. "Aircraft Nuclear Propulsion: An Annotated Bibliography." Prepared for the United States Air Force, History and Museums Program. http://www.afhso.af.mil/ shared/media/document/AFD-141014-032.pdf, accessed 12 January 2013.

Stacy, Susan M. 2000. *Proving the Principle: A History of the Idaho National Engineering and Environmental Laboratory.* U.S. Department of Energy, Idaho Operations Office, Idaho Falls.

Strack, William C. 1971. *Effect of Two Types of Helium Circulators on the Performance of a Subsonic Nuclear-Powered Airplane.* NASA Technical Memorandum No. X–2237.

Tanik Construction and Fairbanks Environmental Services. 2014. "Final 2014 Remedial Action Report, Project Chariot, Cape Thompson, Alaska." www.lm.doe.gov/Chariot/2014RAR_CHT. pdf, accessed 2 January 2017.

Tann, Jennifer (editor). 1981. *The Selected Papers of Boulton & Watt, Vol. 1: The Engine Partnership, 1775–1825.* MIT Press, Cambridge, MA.

Tate, Nick. 1989. "MIT Bombshell Knocks Fusion 'Breakthrough' Cold." *Boston Herald*, 1 May, p. 1.

Taubes, Gary. 1993. *Bad Science: The Short Life and Weird Times of Cold Fusion*. Random House, New York.

Tesla, Nikola. 1888. "A New System of Alternate Current Motors and Transformers." *Transactions of the American Institute of Electrical Engineers, Vol. 5*, pp. 308–327.

Tesla, Nikola. 1891a. "Experiments with Alternating Currents of High Frequency." *Electrical Engineer*, 18 March, pp. 336–337.

Tesla, Nikola. 1891b. "Experiments with Alternate Currents of Very High Frequency and Their Application to Methods of Artificial Illumination." *Transactions of the American Institute of Electrical Engineers, Vol. 17*, pp. 267–319.

Tesla, Nikola. 1891c. "Phenomena of Alternating Currents of Very High Frequency." *Electrical World*, 21 February, pp. 128-130.

Tesla, Nikola. 1892. "Tesla at the Royal Institution." *Scientific American*, 12 March, p. 168.

Tesla, Nikola. 1897. "Tesla on Electricity." *Electrical Review*, 27 January 1897, pp. 46–47.

Tesla, Nikola. 1898. "Tesla Describes His Efforts in Various Fields of Work." *Electrical Review*, 30 November, pp. 344–345.

Tesla, Nikola. 1900. "The Problem of Increasing Human Energy." *Century Illustrated Monthly Magazine*, June, pp. 175–211.

Tesla, Nikola. 1904. "The Transmission of Electric Energy Without Wires." *Electrical World and Engineer*, 5 March, pp. 429–431.

Tesla, Nikola. 1905. "The Transmission of Electrical Energy Without Wires as a Means for Furthering Peace." *The Electrical World and Engineer*, 7 January, pp. 21–24.

Thurston, Robert H. 1887. *A History of the Growth of the Steam-Engine* (4th edition). Kegan Paul, Trench, London.

Thurston, Robert H. 1891. *A Manual of the Steam Engine: For Engineers and Technical Schools; Advanced Courses*. Wiley, New York.

Tierney, John. 1982. "The A-Plane: The $1,000,000,000 Nuclear Bird That Never Flew." *Science* 82, January–February, pp. 46–55.

Tisdale, F. S. 1930. "This House Is Built Like a Tree." *Nation's Business*, February, pp. 42–44, 216–217.

Trubshaw, Brian. 2000. *Concorde: The Inside Story*. Sutton Publishing, Gloucestershire, UK.

Tsyganov, E. N. 2012. "Cold Nuclear Fusion." *Physics of Atomic Nuclei* 75:153–159.

Tucker, Todd. 2009. *Atomic America: How a Deadly Explosion and a Feared Admiral Changed the Course of Nuclear History*. Free Press, New York.

Turba, Ronald D. 2011. *Design of a Nuclear Propulsion System for an Unmanned Aerial Vehicle*. M.S. Thesis, Department of Mechanical Engineering, Vanderbilt University.

Upham, Samuel C. 1878. *Notes of a Voyage to California via Cape Horn: Together with Scenes in El Dorado, in the Years 1849–'50*. Privately printed, Philadelphia.

U.S. Atomic Energy Commission. 1960. "Bioenvironmental Features of the Ogotoruk Creek Area, Cape Thompson, Alaska." Division of Biology and Medicine, Atomic Energy Commission, Washington, DC. www.lm.doe.gov/Chariot/CHT000011.pdf, accessed 30 December 2016.

U.S. Department of Energy. 1989. *Cold Fusion Research: A Report of the Energy Research Advisory Board*. U.S. Department of Energy, Washington, DC.

U.S. Department of Energy. 2004. "Report of the Review of the Low Energy Nuclear Reactions." http://newenergytimes.com/v2/government/DOE2004/DOE-CF-Final-120104.pdf, accessed 29 August 2016.

U.S. Department of Energy. 2005. "Historical American Engineering Record: Idaho National Engineering and Environmental Laboratory – Test Area North, HAER #ID–33–E." U.S. Department of Energy, Idaho Falls, Idaho. https://inldigitallibrary.inl.gov/sti/3028280.pdf, accessed 1 February 2016.

U.S. Department of Energy. n.d. "Plowshare Program." https://www.osti.gov/opennet/reports/plowshar.pdf, accessed 28 December 2016.

Vander Weyde, P. H. 1886. "Some Early Experiences with Electric Motors." *Electrical World* 7:234.

Vaughan, Adrian. 1991. *Isambard Kingdom Brunel: Engineering Knight-Errant*. John Murray, London.

von Guericke, Otto. [1672] 1994. *The New (So-Called) Magdeburg Experiments of Otto von Guericke.* Translated with a preface by Margaret Glover Foley Ames. Kluwer Academic, Dordrecht, The Netherlands.

Walker, William H., and James M. Skibo (editors). 2015. *Explorations in Behavioral Archaeology.* University of Utah Press, Salt Lake City.

Wang, Zuoyue. 2008. *In Sputnik's Shadow: The President's Science Advisory Committee and Cold War America.* Rutgers University Press, New Brunswick, NJ.

Weld, Stuart F. 1887. "Inventions at Panama." *Popular Science Monthly*, December, pp. 145–165.

Wells, Douglas P., et al. 2014. *Low Energy Nuclear Reaction Aircraft – 2013 ARMD Seedling Fund Phase I Project.* NASA Technical Memorandum No. 218283.

Wendt, Gerald. 1951. "A Scientist Previews the First Atomic Airplane." *Popular Science*, October, pp. 98–102.

Wendt, Gerald, and Donald Porter Geddes. 1945. *The Atomic Age Opens.* World Publishing, Cleveland, OH.

White, Lawrence. 1971. *The Automobile Industry Since 1945.* Harvard University Press, Cambridge, MA.

Wilimovsky, Norman J., and John N. Wolfe (editors). 1966. *The Environment of the Cape Thompson Region, Alaska.* U.S. Atomic Energy Commission, Washington, DC.

Witkin, Richard. 1970. "Proxmire Scores Support for SST." *New York Times*, 14 September, p. 77.

Witkin, Richard. 1973a. "2 Airlines Cancel Concorde Orders." *New York Times*, 1 February, p. 1.

Witkin, Richard. 1973b. "Concorde: Profit Is Not the Sole Question." *New York Times*, 7 February, p. 47.

Wong, Sarah. 2000. "The Human Cost of Building the Panama Canal." In *Proceedings of the Ninth Annual History of Medicine Days*, edited by W. A. Whitelaw, pp. 229–232. University of Calgary, Faculty of Medicine.

Zala, Frances F. 1934. "Dymaxion House, Free from Dust, Toil, May Be Anchored Anywhere." *Washington Post*, 7 October, p. B3.

INDEX

A

Abrams M-1 tank, 213

Adelaide (steamship), 78, 79

Adoption (of invention), 268

Aeolipile, 14, 15*f*

Air France, 231, 232

Air pump, 14, 16

Air travel, supersonic airliner, 219–230

Air University Quarterly Review, 170

Airbus, 232–233, 235

Aircraft Nuclear Propulsion program (ANP), 162–164, 165, 167, 168, 170, 171, 172

Airflow cars, 200

Alaska Conservation Society, 189

Albrecht, Ulrich, 171

Alley, Dickenson, 127

Allison Division (General Motors), 199

Alternating current (AC), 6, 115, 118, 119, 120, 121–122, 124, 132

American Journal of Science and Arts, 57

American Pavilion at Expo 67, 148

American Scientist, 250

Amery, Julian, 220

Ampère, André Marie, 56

Antonov An-225 (airplane), 171–172

Archaeology, role of, 2–3

Arc-lighting system, 119

Army Signal Corps, 145

Aspinwall, William, 96

Astor, John Jacob, IV, 126

Atmore, Henry, 50

Atmospheric railway, 33–53, 40*f*, 49*f*, 260, 264, 270

Atomic bomb, 158, 177, 178–179, 181, 239, 240, 241

Atomic Energy Act, 178

Atomic Energy Commission
(AEC), 161, 163, 165, 167, 170,
171, 178, 179–181, 182, 184, 186,
187, 188, 189, 190, 191, 192, 266,
270
"Atoms for Peace," 179, 181, 271
Australian Royal Mail Steamship
Company, 78
Automobile
Airflow cars, 200
author's research on electric
automobiles, 1–2
Chrysler Turbine, 206–213, 206*f*,
215–216, 266, 270, 271
DeLorean DMC-12, 259
Dymaxion car, 140–143, 144*f*,
147, 151, 266–267
as first attempt to apply steam
power in transportation, 13
futuristic, 199
jet-powered, 198, 213, 266
standards for, 213
steam-powered automobile
(French), 23
Axial engine, 61–62, 63, 64, 65, 68

B
B-36H (bomber), 168
B-52 Stratofortress, 164, 169
Baldwin, Jay, 146
Baldwin, Peter, 224
Bangerter, Norman, 249
Barlow, Peter, 42
Barwise (house), 146, 151, 153
Bell, Alexander Graham, 117
Benson, Thomas Hart, 62
Biosphère Montréal, 149*f*

Birmingham, Bristol & Thames
Junction Railway Company, 39
Blenkinsop, John, 34
Blücher (locomotive), 34
Boeing 747, 230
Boeing 2707-300, 227, 233, 235
Boeing Company, 164, 220, 226–
227, 233, 268
Bomb
atomic bomb, 158, 177, 178–179,
181, 239, 240, 241
domesticating of, 177–195
hydrogen bomb (H-bomb), 180,
239, 240
Boston Herald, 248
Boulton, Matthew, 22, 23
Boulton & Watt, 48
Brassey, Thomas, 88
Breguet, Abraham-Louis, 71
Brézin, Michel, 24
Brigham Young University (BYU),
242, 243, 246, 249
Bristol 188 (plane), 223
Bristol Filton Airport, 235
British Aircraft Corporation, 224
British Airways, 231, 232
British and American Steam Navi-
gation Company, 75
British Leyland, 199
British railways, early
development of, 33–38
Broadcasting power, 128, 130–131
Brooks, Jim, 188
Brown, Alfred S., 119, 120
Brown, Harold, 181
Brunel, Isambard Kingdom, 33, 38,
44, 45, 46, 47, 48, 49, 50, 51, 71–
85, 84*f*, 91, 260, 264, 269–270

Brunel, Marc, 71
Buchanan, James, 87
Burgess, William Starling, 141, 143
Bush, Donald J., 141
Butler Brothers, 145

C
Calley, John, 18, 22
Callis, Clayton, 247
Canal Authority of Panama, 112
Cape Thompson, Alaska, 183
Caravelle (airliner), 221
Carlson, W. Bernard, 117, 127, 131
Cars. *See* automobile
Carson, Rachel, 172
Carter, Jimmy, 192
Central Air Force Museum of
    Russia, 235
Century Magazine, 128
Cerf, Alain, 27–29, 30
Chauncey, Henry, 96
Chesebrough, Harry E., 212
Chrysler Corporation, 198–218,
    265, 266, 270
Chrysler Imperial, 214
Chrysler Turbine, 206–213, 206*f*,
    215–216, 266, 270, 271
Chrysler's fourth-generation tur-
    bine engine, 207*f*
Claxton, Christopher, 74
Clegg, Samuel, 39–41, 50
Cold fusion, 241, 242–251, 253,
    260
Columbus (steamship), 95
Comet (jetliner), 220
Commoner, Barry, 189
Commutators, 57, 68, 117, 118
Compagnie Universelle du Canal

Interocéanique de Panama, 104,
    108
Compagnie Universelle du Canal
    Maritime de Suez, 100
Compound magnet and electro-
    tome, 61
Concorde, 219–230, 225*f*, 261, 268,
    270, 271
Condensing steam engine, 16–18,
    17*f*, 19*f*
Congress
    Bayh–Dole Act, 242
    car standards, 213
    as supporting experimental elec-
        trical inventions, 62–63
Conservatoire National des Arts et
    Métiers (National Conservatory
    of Arts and Crafts), 27, 28
Consolidated Vultee Aircraft
    Company (Convair), 162
Cook, Ransom, 59, 60, 267
Cooper, Peter, 66
Croydon railway line, 7, 46, 47, 48,
    51
Crystal House, 143
Cubitt, William, 44, 47, 51, 264
Cugnot, Nicolas-Joseph, 13, 23,
    24–27, 28, 29, 263
Culebra Cut (Panama Canal), 109,
    110*f*, 111, 267

D
Davenport, Oliver, 58
Davenport, Thomas, 57–60, 67, 68,
    267
Davis, Daniel, Jr., 61, 62, 68
Davis, Jefferson, 63
DC-8 passenger jet, 220

DC-10 passenger jet, 230
de Bélidor, Bernard Forest, 20–21
de Courcel, Geoffroy, 220
de Forest, Lee, 132
de Lesseps, Ferdinand, 99–109,
    110–111, 262, 267, 270–271
DeChiaro, Louis F., 252
DeLorean DMC-12 (automobile),
    259
DEMOs, 255–256
Derby, Lord, 84*f*
Desaguliers, John Theophilus, 20
Deutsche Bahn Museum, 28
Development (of invention)
    described, 262
    failures in, 262–266
    a.k.a. research and development
        (R & D), 262
Didcot Railway Centre Museum,
    52
Digester, 16
Dingler, Jules, 106
Direct current (DC), 57, 61, 117,
    121, 243
"Direct cycle" nuclear engine, 265
Dodge Truck, 204
Douglas, 220
Dublin & Kingstown Railway, 42,
    47
Dublin Evening Mail, 43
Dymaxion bathroom, 143, 147
Dymaxion car, 140–143, 144*f*, 147,
    151, 266–267
Dymaxion Corporation, 141
Dymaxion deployment unit, 144–
    145, 145*f*, 151
Dymaxion Dwelling Machine
    Company, 146, 151

Dymaxion house, 137–140, 138*f*,
    146–147, 149, 150, 151–152, 153,
    261, 267
Dymaxion Man, 137
Dymaxion World, 137

E
Eastern Steam Navigation
    Company, 83, 270
Edison, Thomas Alva, 116, 117,
    119, 121
Eiffel, Gustav, 108
Einstein, Albert, 137
Eisenhower, Dwight D., 170, 179,
    180
Electric automobiles, author's
    research on, 1–2
Electric locomotive, 7, 58, 64, 64*f*,
    68, 264
Electric motor
    as example of technology
        rejected by manufacturers in
        one era becoming
        commercialized later, 267
    failure of, 67
    invented by Davenport, 58–59,
        60*f*, 68–69
    invented by Tesla, 118*f*, 120*f*
    invention of, 2
    as most indispensable technology
        of modern world, 55
    negative assessments of, by sci-
        entific authorities, 66–67, 267
    as operating uneconomically
        compared with steam engine,
        65
Electric teeter-totter, 56–57, 57*f*
Electrical World and Engineer, 130

Electro-Magnet and Mechanics'
  Intelligencer, 60
Electro-Magnetic Association, 59,
  60
Electromagnetic gun, 62
Electromagnetism
  discovery of, 55–56
  Henry's use of, 2
  as Page's first love, 61
  power of, 58
Electrotome, 61
Eliot, George Fielding, 178
Emmerson, George S., 80
Emotive functions, 5
Engel, Elwood, 206*f*
Eniwetok Atoll, 240, 255
Environmental movement, growth
  of, 189
Ether, 123

F
F-86 Sabre (fighter), 197
Fairey Aviation Company, 223
Fairey Delta 2 (plane), 223
Falcon (steamer), 95
Faraday, Michael, 2, 122–123
Faraday (cable ship), 89
Fardier (a.k.a. *fardier à vapeur*),
  steam-powered dray, 13, 23–27,
  25*f*, 28, 263, 264
Farmer, Robert C., 214
Fermi, Enrico, 157–158
Fessenden, Reginald, 132
Field, Cyrus, 88, 90
Fifty Car Program, 208
Financing, for promising ideas, 23
Firebird I (futuristic car), 199, 199*f*

Fleischmann, Martin, 242, 243–
  249, 244*f*, 251, 260
Fleming, J. Ambrose, 130
Foote, Don, 186, 187
Fortune magazine, 146
Foster, Norman, 151
4 D house, 137
Fuller, R. Buckminster, 135–155,
  138*f*, 150*f*, 261, 266
Functions, of technology, 4–5
Fusion
  cold fusion, 241, 242–251, 253,
    260
  hot fusion, 10, 241, 246, 247, 253
  nuclear fusion, 180, 240, 245,
    248, 272
Fusion bomb, 239
Fusion experiments, 246, 247, 248,
  249, 255
Fusion machines, 241, 255
Futuristic cars, 199

G
Galileo, 14
Galley, Robert, 230
Gaty, Jack, 146
Geach, Charles, 79, 80
General Dynamics, 164
General Electric (GE), 162, 163,
  165, 167, 172, 226, 227, 260, 264
General Motors, 198, 199–200
Geodesic dome, 6, 135, 147–148,
  151
Geodesics (company), 148
"Geographical engineering," 181,
  190, 270
Ghia, 206, 207
Gibson, John T., 203

Glancey, Jonathan, 223
Gooch, Daniel, 73, 88
Gorbachev, Mikhail, 253
Graham, William L., 151–152
Great Britain (steamship), 76–77, 77*f*, 91
Great Eastern (steamship), 79–90, 81*f*, 82*f*, 86*f*, 260, 269, 270, 271
Great Eastern Steamship Company, 88
Great Ship Company, 85, 88, 90, 269
Great Western Railway (GWR), 38, 72–73, 269
Great Western (steamship), 74–76, 75*f*, 85
Great Western Steamship Company (GWSC), 74, 76, 77
Gribeauval, Jean-Baptiste Vaquette de, 24, 26
Guericke, Otto von, 14
Guppy, Thomas, 73, 74

H
Hadfield, Charles, 50
Hagelstein, Peter, 247, 251
Handley Page, 223
Hanford Reservation, 178
Hawes, William, 78, 85
He 178 (jet), 197
Heat transfer reactor experiments (HTRE), 165–166, 166*f*, 167, 173
Henry, Joseph, 2, 56, 59, 63, 65, 66, 68, 261, 267
Henry Ford Museum, 152*f*
Heron of Alexandria, 14, 15

Hertz, Heinrich, 122
Hertzian waves, 122, 124, 125, 131–132, 265
Hot fusion, 10, 241, 246, 247, 253
HP.115 (plane), 223
Huebner, George J., Jr., 200–201, 203, 204, 205, 208, 210–211, 212, 215, 265, 270
Hughes, G. W., 96
Hughes, Thomas Park, 1
Huizenga, John R., 250
Hunt, Robert, 65–66
Hunter, Celia, 189
Huskisson, William, 37
Huygens, Christian, 16
Hydrogen bomb (H-bomb), 180, 239, 240

I
Ideo-functions, 5
Illustrated London News, 44, 83
Induction coil, 61, 122, 124
Intercity railways, 35, 37
Intercontinental ballistic missiles, 164, 169, 264
International Thermonuclear Experimental Reactor (ITER), 253–255, 254*f*, 255*f*, 262, 267, 271, 272
Inventions
   adoption of, 268
   development of, 262–266
   failure of, 262
   judging feasibility of, 261
   manufacture of, 266–267
   originators of, 5, 260, 261, 269

origins of, 260
promoters of, 3, 5, 33, 34, 35, 59,
    88, 158, 259, 260, 261, 262,
    263, 267, 268, 269, 270
sponsors of, 251, 259–260, 261,
    262, 264, 266, 269, 271
use of, 268–269
Ismail, Viceroy, 100, 102
Ivy Mike. *See* "Super."

J
J47 engine, 167
Jacobi, Moritz, 57
James Watt & Company, 79
Japanese Aerospace Exploration
    Agency (JAXA), 233, 234
Jet planes, 197–198
Jet-powered cars, 198, 213, 266
Johnson, Albert, 188
Johnson, Gerald, 181
Johnson, Robert Underwood, 127
Jones, Steven E., 242
*Journal of Electroanalytical
    Chemistry*, 243, 245

K
Kaiser, Henry, 148
Kalitinsky, Andrew, 158, 159, 260
Keck, George, 143
Keirn, Donald J., 162, 170, 260
Kennedy, John F., 170, 171, 226,
    264
Kistiakowsky, George B., 170

L
Lane Motor Museum, 151
Langer, Lawrence M., 179

Lawrence, Ernest O., 180
Lawrence Livermore National
    Laboratory, 192
Le Bourget Air and Space
    Museum, 235
Lehto, Steve, 209
Leibniz, Gottfried Wilhelm, 242
LeMay, Curtis, 158, 260
Leno, Jay, 216
Leupold, Jacob, 18–20, 23, 24, 25
Lexington report, 161–162
Livermore Lab, 180, 181, 182–184,
    186, 187, 190–191, 192, 265, 270
Liverpool & Manchester Railway,
    35–36, 37
Liverpool Maritime Museum, 90–
    91
Lockheed, 173, 226
Lockheed Martin, 234
Locomotion (locomotive), 35
Locomotive
    Blücher, 34
    electric locomotive, 7, 58, 64, 64*f*,
        68, 264
    invented by Page, 63–65, 64*f*
    Locomotion, 35
    Rocket, 36–37, 36*f*
London & Croydon Railway, 7,
    44, 47
London Globe, 101
Los Alamos National Laboratory,
    178, 180, 181, 192, 239, 241
Los Angeles Times, 203–204
Low energy nuclear reactions
    (LENR), 251–252
Lucas Aerospace, 199

Lucky Dragon incident, 180, 187
Lusitania (passenger liner), 89
Lyman, Arthur, 148

M
MacLaren, Malcolm, 68
MacLeish, Archibald, 140
Madigan, La Verne, 190
Magaziner, Ira C., 249
Magnetic power, 58, 66
Manchester Airport Viewing Park, 235
Manhattan Project, 158, 159, 162, 170, 177, 178, 180, 239
manufacture (of inventions)
  described, 266
  failures in, 266–267
  a.k.a. commercialization, 266
Marconi, Guglielmo, 117, 124–126, 129, 130, 132
Maudslay Sons & Field, 74
Mauretania (passenger liner), 89
McCullough, David, 105, 111
McDonnell Douglas, 268
McDonnell Douglas DC-10, 230
McNamara, Robert S., 170, 171
Mercedes–Benz, 215
Middleton Colliery Railway, 34
Miller, David F., 210, 211–212
Mirage III (fighter), 221
Morgan, J. P., 128, 130, 131
Morgan, Morien Bedford, 220
Morse, Samuel F. B., 62
Morse code, 129
Morton, Henry, 61
Murray and Wood, 34
Museo del Canal Interoceánico de

Panamá (Panama Canal Museum), 111–112
Museum of Civil Aviation in Ulyanovsk, 235
Museum of Flight, 235
Museum of Samara State Aerospace University, 235
Mutual assured destruction (MAD), 169

N
N. M. Rothschild & Sons, 100
National Aeronautics and Space Administration (NASA), 172, 229, 233, 234
National Automobile Museum, 151
National Cold Fusion Institute, 249, 251
National Environmental Policy Act of 1969, 191
National Museum of American History, 68, 216
National Museum of Flight, 235
National Reactor Testing Station, 165, 173
Nature, 243, 246, 247, 248
Nautilus (submarine), 164, 180
Nevada Test Site, 192
New York Herald, 124
New York Herald Tribune, 178
New York Times, 115, 123, 130, 159, 212, 229, 247
Newcomen, Thomas, 18, 22, 24
News Bulletin, 189
Newton, Isaac, 242
Nixon, Richard M., 227

Noguchi, Isamu, 141
North American Aviation, 197
Nuclear bomber/nuclear-powered bomber project, 3, 4, 157–175, 261, 262, 263–264, 270, 271. *See also* NX2 (Convair plane)
Nuclear Energy for Propulsion of Aircraft (NEPA), 158, 159, 161
Nuclear excavation, 181–183
Nuclear fallout, 161, 180, 182, 183, 184, 186, 187, 188, 189
Nuclear fusion, 180, 240, 245, 248, 272. *See also* cold fusion; hot fusion
Nuclear Information, 189
Nuclear power plant, 158, 164, 180
Nuclear submarine, 157, 162, 164, 169, 171, 180
Nuclear war, 169, 182
Nuclear weapons, 162, 164, 178, 180, 192
NX2 (Convair plane), 168, 169*f*, 172

O
Oak Ridge National Laboratory, 158, 162, 163, 178, 241
Oersted, Hans Christian, 55–56
Ogotoruk Creek, 183, 184, 186, 187, 188, 193*f*, 265
Ohio (steamship), 58, 96
O'Neill, Dan, 186–187, 188, 190
O'Neill, John T., 179
Operating Manual for Spaceship Earth (Fuller), 148
Operation Ivy, 240
Oppenheimer, J. Robert, 159

Organization of Petroleum Exporting Countries (OPEC), 230
Originators (of inventions), 5, 261, 262, 269
Osceola (sailing ship), 95
Oscillating transformer, 122, 123, 125, 126
Otis, Fessenden, 98

P
P-1 reactor, 163
Pacific Mail Steamship Company, 96
Paddle wheel engine (*Great Eastern*), 82*f*
Page, Charles Grafton, 58, 61, 62, 63–65, 67, 68, 122, 264
Pan Am, 230
Panama Canal, 104–112, 106*f*, 110*f*, 262, 267, 270–271
Panama Railroad Company, 96–98
Papin, Denis, 16, 22
Park, Robert L., 250
Parker, Ronald R., 248
Pasha, Saïd, 99, 100
Pathological science, 250–251
The Path Between the Seas: The Creation of the Panama Canal 1870–1914 (McCullough), 105
Patterson, William, 74
Patty, Ernest, 186
Pauling, Linus, 186
Pearson, Philip, 141, 143
Pease, Edward, 34–35
Peck, Charles F., 119, 120, 121
Pender, John, 88

Pim, James, Jr., 42
Pinkus, Henry, 38, 39
Piston carriage, 39, 40, 50
Piston engine, 197, 198, 200, 201, 212, 214–215, 219
Playboy magazine, 149
Plutonium-239, 239
Plymouth 1956, 203
Plymouth 1960, 204
Plymouth Belvedere 1954, 201
"Pneumatic" railway, 39
Point Hope Village Council/Point Hope Eskimos, 184, 186, 187, 188, 190, 191, 192, 265
Pons, Stanley, 242–249, 244*f*, 251, 260
Pons–Fleischmann cell, 243–244
Pool, Robert, 247
Popular Mechanics, 261
Potter, Robert D., 179
Pratt & Whitney Aircraft, 163, 172, 260
Prefabricated homes, 137
President's Science Advisory Committee, 170
Pressure cooker, 16
Project Chariot, 183–195, 185*f*, 265–266, 271
Project Plowshare, 182, 189, 190, 192
Promoters (of inventions), 3, 5, 33, 34, 35, 59, 88, 158, 259, 260, 261, 262, 263, 267, 268, 269, 270
Proxmire, William, 226, 227
Pruitt, William, 188, 189
Puskás, Tivadar, 119

R
Radioactive exhaust, 265, 271
Radioactive pollution, 172. *See also* nuclear fallout
Railway Times, 42
Railways. *See also specific railways*
across Isthmus of Panama, 96–98, 97*f*
atmospheric railway, 33–53, 40*f*, 49*f*, 260, 264, 270
as helping to accelerate industrialization and foster growth of middle and upper classes, 98
intercity railways, 35
"pneumatic" railway, 39
steam railways. *See* steam railways
Reagan, Ronald, 253
Reines, Frederick, 181
Rennie & Maudslay, 48
Resources, consequences of lack of, 262–264
Rickover, Hyman G., 164
Righi, Augusto, 124
Ritchie, Edward, 61
Robert Stephenson & Company, 35, 36, 37, 38
Rocket (locomotive), 36–37, 36*f*
Roebuck, John, 22, 23
Roentgen, Wilhelm, 117
Rolls-Royce/Snecma Olympus 593 engine, 222
Rousseau, Denis, 250–251, 252
Rover, 198–199
Russell, John Scott, 78, 79, 80, 83, 84*f*
Russian MiG-15 (fighter), 197

S
Saerchinger, César, 178
Salamanca, 34
Samuda, Jacob, 39–41, 44, 50, 260
Samuda, Joseph, 39, 41, 44, 260
Saturday Evening Post, 208
Savery, Thomas, 16–18, 22
Science, 189, 247
Science News Letter, 189
Scientific American, 103
Screw engine (*Great Eastern*), 82*f*
Sea–Air–Space Museum, 235
Sears, Roebuck & Company, 137
707 passenger jet, 220
747 passenger jet, 230
Shipping Gazette, 103
Sierra Club, 190
Silent Spring (Carson), 172
Silliman, Benjamin, 59
Sirius (ferry), 75–76
Smalley, Orange, 58
Smith, Frederick, 42
Smithsonian Institution, 216
Socio-functions, 5
South Devon railway line, 7, 44, 45, 46, 47, 48, 49*f*, 50, 51, 52, 269, 270
Spectacular Flops
  contributing causes of, 264–266
  life history of, 4
  as often augmenting engineering science, 270
  as potentially harming reputation of promoters, 270
  as special species of failure, 3
Sponsors (of inventions), 251, 259–260, 261, 262, 264, 266, 269, 271

SR-71 (spy plane), 169, 226
Steam dray, 23, 24–25
Steam engine
  Cerf's replica of Cugnot's *fardier*, 29*f*
  invented by Cugnot, 24–27, 25*f*, 26*f*, 28
  invented by Leupold, 18–20, 20*f*, 23, 24, 25
  invented by Newcomen, 18, 19*f*, 20, 21, 24
  invented by Savery, 16–18, 17*f*, 21
  invented by Watt, 21–22, 23
  as powering railroads and machinery, 58
  putting of on carriage or wagon, 34
  replacement of with electric motor, 61
Steam power, applied to transportation, 13
Steam railways, 41, 42, 45, 46, 50–51, 264
Steam technology, history of, 14–30
Steam-powered automobile (French), 23
Steam-powered bucket excavator, 106*f*
Steamships, 58, 71–91, 98, 260–261, 269–270. *See also specific steamships*
Stephens, John L., 96
Stephenson, George, 34, 35, 44, 45
Stephenson, Robert, 73
Stephenson gauge, 38

Stockton & Darlington Railway, 34, 35
Stokowski, Evangeline, 143
Stokowski, Leopold, 143
Strauss, Lewis, 182
Studying Technological Change: A Behavioral Approach (Schiffer), 3
Sturgeon, William, 56
Sud-Aviation (later Aérospatiale), 224
Suez Canal, 99–103, 102*f*
Suez Canal Authority, 111
Supersonic airliner, 219–238, 269
Supersonic Transport Aircraft Committee (STAC), 220–221
Supersonic transport plane (SST), 220, 222, 226, 227, 228, 233, 234, 234*f*
"Super" (nicknamed Ivy Mike), 180, 239, 240, 255
Surrey Iron Railway, 33–34
Swartz, L. Gerard, 188

T
Taking Charge: The Electric Automobile in America (Schiffer), 27
Tampa Bay Automobile Museum, 27, 29
Techno-functions, 4–5
Technological change, study of, 2
Technological failures, spectacular flops as, 3
Technology
    electric motor as most indispensable technology of modern world, 55
    functions of, 4–5
    history of steam technology, 14–30
    life history of, 259
Telegraph, 45, 46, 48, 49, 62, 64, 71, 88, 99, 119, 120, 128, 130
Teller, Edward, 180, 181, 182, 183, 184, 239, 240, 248, 265, 270
Tender, 34
Tesla, Djuka, 117
Tesla, Milutin, 117
Tesla, Nikola, 115–134, 116*f*, 127*f*, 260
Tesla coil, 122, 123, 125, 126, 130
Tesla: Inventor of the Electrical Age (Carlson), 117
Tesla Science Center, 133
Theatri Machinarum Hydraulicarum (*Theater of Hydraulic Machines*), Leupold, 19–20
"Thermoelectric" motor, 119, 120
Thermonuclear device, 239
Time magazine, 116*f*, 148, 208
Tokamak, 241, 253, 254*f*, 255, 255*f*, 262, 267, 271, 272
TorqueFlite, 207
Town Topics, 125–126
Toyota, 251
Transcontinental Railroad, 99
Trick photography, 127*f*
Trubshaw, Brian, 223, 229
Truman, Harry S., 177, 178, 240
Tu-144 supersonic plane, 228*f*, 229, 232, 235, 269
Tupolev Design Bureau, 228
Turbine's variable nozzle system, 205*f*

Turboflite (sports car), 204
Turbojet engine/turbine engine,
  158–159, 160, 162, 165, 166, 169,
  197–218, 202*f*, 227, 263. *See also*
  Chrysler Turbine
Turcat, André, 229
TWA, 230
2707-300 (plane), 227, 233, 235

U
U-2 (spy plane), 169
Udvar–Hazy Center (Smithsonian
  National Air and Space
  Museum), 235
Ulam, Stanislaw, 239, 240
Une Ville Flottante (*A Floating
  City*), Verne, 88
Union Carbide Nuclear Company,
  163
University of California Radiation
  Laboratory, 180
University of Utah, 25, 242, 243,
  245, 246, 247, 248, 249, 251
Upham, Samuel C., 95–96
Uranium-235 (U-235), 179, 180,
  239
U.S. Department of Defense
  (DOD), 161, 162, 163, 170
U.S. Department of Energy
  (DOE), 192, 242, 249, 250, 251,
  252
U.S. Environmental Protection
  Agency (EPA), 212
U.S. Naval Surface Warfare
  Center–Dahlgren Division
  (NAVSEA), 252
U.S. News & World Report, 162

U.S. Patent and Trademark Office,
  261
Use (of inventions)
  failures and accidents during, 269
  as stage of invention following
    adoption, 268–269

V
Vallière, Monsieur, 28, 29
Velikhov, Evgeny, 252
Verne, Jules, 88
Victoria (steamship), 78, 79
Viereck, Leslie, 188
Vlaha, Richard and Patricia, 209
von Braun, Wernher, 164
*Voodoo Science: The Road from
  Foolishness to Fraud* (Park), 250

W
Wakefield, Henry, 84*f*
Warden, James S., 128
Wardenclyffe, 128, 129*f*, 131, 132–
  133, 265
Warren, Shields, 161
Warren, Waldo, 137
Washington Post, 140
Watkins, James D., 249
Watt, James, 21–23
Westinghouse, George, 121
Weston dynamo, 120
Weyde, P. H. Vander, 66
White, Stanford, 128
Wichita house, 152, 152*f*
Wide-gauge railway lines, 38
Wilderness Society, 190
Wireless (radio), 117, 124–125, 129,
  130, 132

Wood, Ginny, 189
Workshop on Cold Fusion, 249
World System (of Tesla), 117,
123–124, 128, 130, 131, 265,
270
Wormwood Scrubs, 39, 41, 42, 43,
44, 48, 260

X
X211 (turbojet), 167

XNJ140E propulsion system, 167–
169, 167*f*
X-rays, 117

Y
Yelyan, Eduard V., 229

Z
Zala, Frances F., 140
Zeder, James C., 201, 203, 204